Nanomedical Brain/Cloud Interface

Explorations and implications

Online at: https://doi.org/10.1088/978-0-7503-2144-0

Nanomedical Brain/Cloud Interface

Explorations and implications

Edited by
Frank J Boehm

Founder: NanoApps Medical, Inc.

and

Co-Founder: NanoApps Athletics, Inc.

IOP Publishing, Bristol, UK

Frank J Boehm has asserted his right to be identified as the editor of this work in accordance with sections 77 and 78 of the Copyright, Designs and Patents Act 1988.

ISBN   978-0-7503-2144-0 (ebook)
ISBN   978-0-7503-2142-6 (print)
ISBN   978-0-7503-2145-7 (myPrint)
ISBN   978-0-7503-2143-3 (mobi)

DOI   10.1088/978-0-7503-2144-0

Version: 20251101

IOP ebooks

British Library Cataloguing-in-Publication Data: A catalogue record for this book is available from the British Library.

Published by IOP Publishing, wholly owned by The Institute of Physics, London

IOP Publishing, No.2 The Distillery, Glassfields, Avon Street, Bristol, BS2 0GR, UK

US Office: IOP Publishing, Inc., 190 North Independence Mall West, Suite 601, Philadelphia, PA 19106, USA

I lovingly dedicate this book to the memory of my father, Josef Boehm, who ignited in me the flame of imagination by the example of his unquenchable curiosity, quest for knowledge, and fascination with life and the universe; to my mother, Charlotte Boehm, whose amazing fortitude, generosity, and contagious enthusiasm for life have inspired all around her to never give up on their dreams and to always look up (we recently celebrated her 99th birthday!); and to my sweeties, Liz Balfour and Jazmyn Balfour–Boehm, whose unconditional love and support have allowed me to realize my dreams. Further, to my amazing grandsons, Skye (SkyeGuy) Barney and Linden Barney (beautiful sons of Jazmyn and Steve Barney) who are incarnations of pure love and joy, for whom I deeply hope and trust that the world they inherit will be one that is far more unified, loving, equitable, wise, and compassionate.

I also dedicate this book to several late colleagues who I will always consider 'forces of nature,' whose presence, passion, diligence, leadership, and kindness will be sorely missed—Veer Endra, Ioan Opris, Ned Seeman, Selwyn Super, and Jack Taunton.

Contents

Preface   xii

Acknowledgments   xv

Editor biography   xvii

List of contributors   xix

**1   Ethics of a human brain/cloud interface and transparent   1-1
     shadowing**
     Melanie Swan, Jeffrey V Rosenfeld and Frank J Boehm

1.1 Introduction   1-1
1.2 Potential ethical considerations for a brain/cloud interface   1-2
     1.2.1 Preamble   1-2
     1.2.2 Ethics of brain-embedded medical neuralnanorobots   1-2
     1.2.3 Ethics of transparent shadowing   1-3
     1.2.4 Warning against an overly cautious ethics   1-3
     1.2.5 Working toward an ethics of immanence   1-4
1.3 Brain/cloud interface—neural trust, blockchain, and neural   1-4
     cryptographic security
     1.3.1 Neural trust   1-4
     1.3.2 Blockchain-based neural trust   1-6
     1.3.3 Billion-neural-nanorobot orchestration and system control   1-8
     1.3.4 Questions, risks, and limitations   1-9
1.4 Brain/cloud interface—potential moral considerations   1-11
     1.4.1 Morality of brain-embedded medical nanorobots   1-11
     1.4.2 Morality of transparent shadowing   1-17
1.5 Potential sociological impacts   1-18
     1.5.1 Sociological impacts of brain-embedded medical nanorobots   1-18
     1.5.2 Sociological impacts of transparent shadowing   1-21
1.6 Philosophical/ontological considerations   1-25
     1.6.1 Philosophical perspectives on brain-embedded medical   1-26
            nanorobots
     1.6.2 Philosophical perspectives on transparent shadowing   1-26
     1.6.3 Impacts on human civilization—TS as a gateway to a   1-26
            ten-billion-synapse world mind?
1.7 Higher planes of existence   1-27
     1.7.1 Brain/cloud-interface-facilitated access to universal   1-27
            energy fields and spiritual realms

1.8   Requesting cognitive quiet   1-27
1.9   Conclusions   1-28
       Appendices   1-29
       References and further reading   1-35

2   Analysis of power, locomotion, communications, and navigation   2-1
**   for microbots in the brain**
        Tad Hogg

2.1   Introduction   2-1
2.2   Chemical power   2-4
2.3   Locomotion   2-7
2.4   Communications   2-9
2.5   Navigation   2-12
2.6   Discussion   2-15
       Acknowledgments   2-17
       References   2-17

3   Combining a neural bypass and a neural allograft to develop a   3-1
**   prosthetic thalamus**
        Miguel Pais-Vieira, António J Salgado and Carla Pais-Vieira

3.1   Introduction   3-2
3.2   Method and system for restoring communication in the central 3-2
       nervous system
3.3   Neural bypass   3-4
3.4   Scaffold formed from microcolumns   3-5
3.5   Remote training of the neural graft   3-6
3.6   Remote control of the neural bypass by a remote neural graft 3-7
3.7   Implantation of the trained neural graft   3-8
3.8   Gradual removal of the neural bypass   3-9
3.9   Conclusions   3-10
       Definitions   3-11
       References and further reading   3-11

4   Neurotech frontiers: ethics, identity, and the evolution of   4-1
**   brain–computer interfaces**
        Ingrid Vasiliu-Feltes

4.1   Introduction   4-2
4.2   Section I: ethical foundations of neurotechnology and brain–computer   4-5
       interfaces

	4.2.1 Digital ethics in neurotechnology: an advanced examination	4-5
	4.2.2 Societal ethics in neurotechnology: a holistic approach	4-7
	4.2.3 Corporate ethics in neurotechnology: a responsible framework	4-9
	4.2.4 Applied ethics in neurotechnology: an ethical imperative	4-10
	4.2.5 Bioethics and medical ethics considerations	4-12
	4.2.6 Brain–computer interface research trends	4-13
4.3	Section II: redefining digital identity in brain–computer interfaces	4-15
	4.3.1 Redefining identity	4-15
	4.3.2 Multiple digital identities	4-17
	4.3.3 Digital identity in brain–computer interfaces	4-19
	4.3.4 Dynamic informed consent in brain–computer interfaces	4-20
4.4	Section III: ethical considerations in hybrid augmented brain–computer interfaces	4-21
	4.4.1 Ethics of hybrid augmented workflows: navigating complex human–digital interactions	4-21
	4.4.2 Tailoring brain–computer interfaces to diverse learning styles and intelligence	4-23
	4.4.3 Impact of brain–computer interfaces on employment	4-24
4.5	Section IV: brain–computer interface harmonization	4-24
	4.5.1 Brain–computer interface cyberethics	4-24
	4.5.2 Sustainable brain–computer interfaces	4-25
	4.5.3 Diverse and inclusive brain–computer interfaces	4-25
4.6	Section V: deep tech convergence	4-25
	4.6.1 Brain–computer interfaces and AI	4-25
	4.6.2 Brain–computer interfaces and digital twins	4-26
	4.6.3 Precision brain–computer interfaces	4-26
	4.6.4 Brain–computer interfaces and multicloud computing	4-27
	4.6.5 Brain–computer interfaces and high-speed networks	4-27
	4.6.6 Brain–computer interfaces and satellite internet	4-27
	4.6.7 Brain–computer interfaces and quantum computing	4-27
4.7	Section VI: strategic considerations	4-28
	4.7.1 Corporate board-level considerations	4-28
	4.7.2 C-suite responsibilities	4-28
	4.7.3 Management-level responsibilities	4-28
	4.7.4 Employee-level considerations	4-28
4.8	Future directions	4-28
4.9	Conclusions: the moral imperative of responsible brain–computer interface deployment	4-29
	References and further reading	4-30

5   Impact of a brain/cloud interface on humanity—gateway   **5-1**
**   to the ten-billion-synapse world mind**
Melanie Swan

5.1   Brain/cloud interface cloudmind   5-2
      5.1.1 Brain/cloud interface technologies   5-2
      5.1.2 Brain/cloud interface, medical nanorobots, and neuralnanorobots 5-3
      5.1.3 Purpose of brain/cloud interfaces: the Kardashev-plus society   5-4
5.2   The brain/cloud interface project: neural signaling and   5-7
      neuralnanorobot instantiation
      5.2.1 Summary of neural signaling   5-7
      5.2.2 Neural cells and neuralnanorobot complements   5-10
      5.2.3 Neurocurrencies   5-12
5.3   Brain/cloud interface hardware: quantum computing for the brain   5-15
      (quantum brain/cloud interface)
      5.3.1 Communication and connectivity platforms   5-15
      5.3.2 Quantum computation   5-16
      5.3.3 Nature's built-in quantum security features   5-19
5.4   Brain/cloud interface software: holographic control theory   5-20
      5.4.1 Holographic correspondence (the anti-de Sitter/conformal field 5-20
             theory correspondence)
      5.4.2 The black hole information paradox   5-22
      5.4.3 The anti-de Sitter/conformal field theory correspondence as a   5-23
             brain/cloud interface control theory
      5.4.4 The anti-de Sitter/conformal field theory correspondence as a   5-24
             control model for complex domains
      5.4.5 Anti-de Sitter/conformal field theory correspondence studies   5-25
5.5   Brain/cloud interface operating software: bioblockchain neuroeconomy   5-26
      5.5.1 Bioblockchain neuroeconomy   5-26
      5.5.2 Tech specs: brain/cloud interface neuralnanorobot network   5-26
             system requirements
5.6   Peak-performance brain/cloud interface cloudminds   5-31
      5.6.1 Instantiating well-formed groups   5-31
      5.6.2 Overcoming barriers to large-scale group collaboration   5-32
      5.6.3 Cloudmind activities: what does the brain/cloud interface   5-37
             cloudmind do?
      5.6.4 Classes of cloudminds   5-39
5.7   Risks and limitations   5-40
5.8   Conclusions   5-42
      Glossary   5-43

Appendix A: Relative sizes of neural entities and neuralnanorobots   5-45

Appendix B: B/CI technical requirements and implementation phases   5-50

References and further reading   5-52

6   The ultimate chip   6-1

Howard Bloom

6.1   Aladdin's lamp   6-1

6.2   Boolean algebra and the ultimate chip   6-2

6.3   Encephalomyelitis/chronic fatigue syndrome   6-2

6.4   Intelligent agents   6-3

6.5   Golden age of radio   6-4

6.6   Personalized Ashley   6-4

6.7   Neuralink   6-5

6.8   The ultimate chip   6-5

References   6-6

Preface

Since the publication of my first book (*Nanomedical Device and Systems Design: Challenges, Possibilities, Visions* CRC Press 2013), it has become increasingly evident that synergies between the concurrently rapid advances in nanotechnology and nanomedicine, in conjunction with the recent exponential progress in artificial intelligence (AI), would likely facilitate the capacity to actualize three envisaged global-scale paradigm-shifting technology platforms. Thus, I endeavored to explore these possibilities through the simultaneous generation of what I consider to be a set of three companion books, based on the rationale that I believe these applications will be inextricably linked.

The first volume in the set (Boehm 2025 *Global Health Care Equivalency in the Age of Nanotechnology, Nanomedicine and Artificial Intelligence* (CRC Press)) investigates the potential emergence of an envisaged equitable, cost-effective, non-monopolizable, globally distributed/decentralized AI/quantum computation (QC)-driven healthcare system, which would operate on safe/secure nested quantum-encrypted blockchains. With the attainment of mature global healthcare equivalency (GHCE), which might be significantly expedited when advanced AI (and soon enough the emergence of artificial general intelligence (AGI) and artificial superintelligence (ASI)) are coupled with QC...

> ...we can envisage a future world where any individual on the planet has access to the same advanced and cost effective nanomedical diagnostic and therapeutic technologies, no matter how wealthy or impoverished they are, no matter where they happen to reside, or under what conditions they live. Progress toward this goal will be incremental, with each successive wave of nanomedical technologies being more advanced than the previous wave. (Frank Boehm, NanoApps Medical, Inc. 2024).

The prerequisite technology platform and tipping point for mature GHCE will arrive with the emergence of the hypothetical (for now) molecular manufacturing (MM); the topic of my second book (*Molecular Manufacturing: The Future of Nanomedicine* (Boehm, CRC Press, ~2025)). It is envisaged that MM may enable the cost-effective domestic fabrication of virtually any consumer item (including nutritious gourmet foods) as well as envisaged autonomous nanomedical devices (multiple components and systems of which are articulated in my first book), which may have the capacity to address all that ails us, including the disease of aging.

This third volume in the series (*Nanomedical Brain/Cloud Interface: Explorations and Implications* (B/CI))...

> ...embarks on an in depth exploration of the future (hypothetical) possibility that the cerebral cortex of the human brain might be seamlessly, safely, and securely connected with the Cloud as a Brain/ Cloud Interface (B/CI). Such an envisaged nanomedically facilitated

cognitive augmentation may consist of a highly integrated network of sophisticated autonomous nanorobotic devices coupled with advanced AI toward the enablement of instantaneous and finely controllable connectivity with the Cloud.

This interface might serve as a personalized conduit through which individuals would not only have immediate access to virtually any facet of cumulative human knowledge, but also the optional and specialized capacity to engage in real time fully immersive experiential/sensory engagement, including what is referred to as 'Transparent Shadowing' (TS), where individuals may experience episodic segments of the lives of other willing participants anywhere on the planet in real time, at full sensory resolution.

Further to a preceding investigation and articulation of the potential technical aspects of a B/CI (Martins *et al* 2019), this book will also delve into its ethical, moral, sociological, legal, and philosophical implications. The prospective usefulness and benefits of such a powerful technology must be tempered by a careful consideration of the perceived risks and potential for misuse, such as neocortex hacking and nefarious thought manipulation and control, toward the formulation of prudent future policies.

Myriad non-trivial questions will be brought to bear toward elucidating how B/CI technologies might potentially impact ones sense of self, and how that self may relate to others, the world, and beyond. What might the benefits, risks, and consequences for human civilization be, when individuals have access to unprecedented opportunities for significant personalized cognitive, sensual, and experiential augmentation, and who may, through the use of B/CI technologies, be so intimately interconnected? (Frank Boehm, NanoApps Medical, Inc. 2024)

As is the case for GHCE, the prerequisite technology platform and tipping point toward a mature B/CI will arrive with the advent of MM, which may facilitate the cost-effective domestic fabrication of the advanced autonomous neuralnanorobots required to enable it (Martins *et al* 2019). (*Note: It is acknowledged that concurrently with the development of nanomedical neuralnanorobotics, it is likely that rapid progress in AI and the subsequent potential near-term emergence of AGI and ASI may facilitate the conceptualization and development of completely noninvasive B/CIs via the sophisticated manipulation of energy, frequency, and vibration, per the visionary concepts of Nikola Tesla*) (Tesla 1892). A further essential requirement, prior to the programmed self-installation of specialized B/CI neuralnanorobots, relates to our individual uniqueness, which extends to the distinct physiological cellular compositions of our brains. Consequently, the critical initial step of brain mapping at ultrahigh cellular/spatial resolution will be an important standard B/CI procedure. A conceptual nanomedical strategy through which this might be accomplished is discussed in an earlier chapter that I co-authored with Dr Angelika Domschke (Domschke and Boehm 2017).

As described in the first two books in this series, a mature B/CI may, in effect, facilitate even greater enhancements, efficacies, and optimizations for MM and GHCE, which, from the author's perspective, would close the loop. The reason for this is that every one of these three technology platforms might synergistically cross-pollinate, with each possessing the innate capacity to reciprocate. Further downstream, the broad tech-trusted acceptance and ubiquitous implementation of a mature B/CI may be nothing less than the next evolutionary step for humanity, one where we seamlessly merge with our technologies. This may come to be realized as a fail-safe contingency that supports humanity in keeping pace with AI/AGI/ASI and engages us in a continuum of perceived relevance while serving as a robust force against the prospect of human redundancy (or worse) as an unintended consequence of exponentially advancing recursive AI/AGI/ASI.

References

Domschke A and Boehm F J 2017 Application of a conceptual nanomedical platform to facilitate the mapping of the human brain: survey of cognitive functions and implications *The Physics of the Mind and Brain Disorders* (Springer Series in Cognitive and Neural Systems Series in Cognitive and Neural Systems) (Cham: Springer) **11** 741–71

Martins N R B *et al* 2019 Human brain/cloud interface *Front. Neurosci.* **13** 112

Tesla N 1892 Experiments with alternate currents of high potential and high frequency *Electr. Eng.* **10** 1–12

Acknowledgments

I am indeed deeply grateful to all the contributing authors and co-authors listed in the Contributors section for their incredible efforts in creating this book and infusing their amazing and inspiring intellects and life forces into it. I would surely be remiss if I did not acknowledge K Eric Drexler for presenting to the world his original vision of the boundless possibilities of nanotechnology, molecular manufacturing, and nanomedicine, as articulated in his highly inspirational book, *Engines of Creation* (1987). Furthermore, my true appreciation is extended to Robert A Freitas Jr for further enlightening me as to the virtually limitless and exciting potential of nano-medicine through his acute insights, most excellent and groundbreaking book, *Nanomedicine, Volume I: Basic Capabilities* (1999), and *Nanomedicine* book series. Also, my sincere gratitude goes to Richard Satava (former Defense Advanced Research Projects Agency (DARPA) program manager and professor of surgery at the University of Washington, Seattle) for initially inspiring me to articulate my nanomedical concepts on paper while advising me to always endeavor to amply support (via the scientific literature) their potential viability and promise. Finally, my special thanks and deep gratitude go to Amanda Scott (CEO - NanoApps Medical, Inc. and Creative Director - Alias Studio) for her amazing and tireless efforts in support and promotion of my nanomedical concepts and publications, as well as the latest progress in nanotechnology, nanomedicine, and AI through the creation and curation of our excellent NanoApps Medical, Inc. and NanoApps Athletics, Inc. sites and social media. I also give my sincere thanks and kudos to Heinz Hoenen for his exceptional efforts and contributions to our international social media presence.

I would like to express my deep appreciation to the following individuals for providing information, insights, and other forms of support in facilitating the realization of this project: Amara Angelica, Elizabeth Balfour, Jazmyn Balfour–Boehm, Steve Barney, Aicheng Chen, Han Chen, Angelika Domschke, Ted Duke, Claude Giroux, Urs Hafeli, David Harakal, Robert R Hieronimus, Jonathan Kitzen, Haijin Liu, Adriana Marias, Kannan Mavila, Margaret Morris, Sabino Padilla, Shannon Smith, Frank Soda, Melanie Swan, and Jack Taunton.

I sincerely apologize to any individuals whom I may have inadvertently overlooked.

My sincere thanks and gratitude go out to my publisher (Institute of Physics Publishing (IOPP)) for having the confidence and trust in me to see this project through and for allowing me complete freedom to explore some of the exciting possibilities of an envisaged B/CI with its potentially significant benefits for humanity and the planet. In particular, I wish to convey my appreciation to Michael Slaughter, Isabelle Defillion, Erika Radzvilaite, and Bethany Hext for their incredible patience and support throughout the writing process. In addition, I very much appreciate the efforts of Bethany Hext and the staff at IOPP in the production of this book.

Finally, I will forever be grateful to my truly inspiring and wise father, Josef Boehm, and my ever-sweet and generous mother, Charlotte Boehm. I also thank my loving sister, Renata Swanson; my brother, John Boehm; and Elizabeth Balfour. I am deeply thankful to our incredible daughter Jazmyn Balfour–Boehm and her husband Steve Barney, who blessed us with our beautiful and amazing grandsons Skye (SkyeGuy) Barney and Linden Barney. Their unconditional love, encouragement, and unwavering support have made this book possible.

Editor biography

Frank J Boehm

Frank J Boehm comes from a diverse background that includes a 12-year stint as a full-time rock/pop musician in various bands that toured North America (1975–87) and 17 years as a senior mechanical designer involved in the design and development of dedicated robotic tooling for complex automated assembly systems (1987–2004). He has been further engaged as a nanotechnology consultant, technology scout, product designer, research associate, and academic writer/editor for several international corporations and university laboratories since 1996. Frank has been involved with nanotechnology, especially nanomedicine, since 1996 (autodidactically trained), which inspired the development of numerous concepts and designs for advanced nanomedical diagnostic and therapeutic components, devices, and systems to potentially address myriad disease states. His aim is to develop and transform these concepts into real-world applications for global benefit. Frank serendipitously encountered the concept of nanotechnology on the internet and immediately became fascinated with its virtually limitless potential, particularly as it relates to the field of medicine. He passionately proceeded to evolve and textually articulate various advanced near-term and longer-term nanomedical concepts and designs. Concomitantly, he initiated correspondence with numerous nanotechnology and nanomedicine research scientists and thought leaders from across the globe and founded a startup company, NanoApps Medical, Inc. (2009), with the aim of investigating and developing advanced, innovative, and cost-effective nanomedical diagnostic and therapeutic devices and systems for the benefit of humanity. Upon recognizing the immense potential of nanomedicine to impart positive paradigm shifts across the medical domain (e.g. precisely targeted drug delivery, vascular/neurological/cellular plaque removal, eradication of cancers via hypothermic nanoparticles, completely noninvasive surgical procedures, the enhancement of physiological systems, and extended longevity), Frank was deeply motivated to write more extensively on the topic. This culminated in the authoring/publication of his first book, *Nanomedical Device and Systems Design: Challenges, Possibilities, Visions* (CRC Press, 2013). Subsequently, he authored/co-authored numerous book chapters and journal articles, including an in-depth paper entitled 'Human Brain/Cloud Interface' (*Frontiers in Neuroscience*, 2019), and co-founded NanoApps Athletics, Inc. in the same year. Frank then set his focus on a far more expansive investigation into the possibilities of nanotechnology/nanomedicine coupled with AI/AGI/ASI and quantum computation via the simultaneous compilation of a set of three companion books (with multiple chapter contributors for each) that explore a trifecta of envisaged global-scale paradigm-shifting technology platforms: *Global Health Care Equivalency in the Age of Nanotechnology, Nanomedicine and Artificial Intelligence* (CRC Press, 2025), *Molecular Manufacturing: The Future of Nanomedicine* (CRC Press, 2025), and *Nanomedical Brain/Cloud Interface: Explorations and Implications* (IOP Publishing, 2025).

List of contributors

Howard Bloom
Author
The Lucifer Principle and The Case of the Sexual Cosmos: Everything You Know About Nature Is Wrong.

Frank J Boehm
NanoApps Medical Inc.
NanoApps Athletics Inc.
NanoApps Consulting
Vancouver, British Columbia, Canada
Thunder Bay, Ontario, Canada

Tad Hogg
Institute for Molecular Manufacturing, Palo Alto, CA, USA

Carla Pais-Vieira
Universidade de Aveiro
Brain-Machine Interface Research Laboratory, Centro de Investigação
Interdisciplinar em Saúde (CIIS), Instituto de Ciências da Saúde (ICS),
Universidade Católica, Portuguesa, Porto,
Portugal

Miguel Pais-Vieira
Institute of Biomedicine (iBiMED), Department of Medical Sciences, Universidade
de Aveiro, Aveiro, Portugal

Jeffrey V Rosenfeld
Senior Neurosurgeon at The Alfred Hospital, Melbourne, Australia

António J Salgado
Life and Health Sciences Research Institute (ICVS), School of Medicine, University
of Minho, Campus de Gualtar, Braga, Portugal

Melanie Swan
UCL Centre for Blockchain Technologies, London, UK
and
DIYgenomics, New York, USA

Ingrid Vasiliu-Feltes
University of Miami, Miami, FL, USA

Chapter 1

Ethics of a human brain/cloud interface and transparent shadowing

Melanie Swan, Jeffrey V Rosenfeld and Frank J Boehm

This chapter serves as a companion to our earlier paper, 'Human brain/cloud interface' (Martins *et al* 2019), which investigates the ethical aspects of a hypothetical (for now) nanomedically enabled brain/cloud Interface (B/CI). This will facilitate discussions on its prospective usefulness and benefits and allow the reader to explore the risks and potential for misuse, thereby supporting the formulation of a prudent regulatory framework. There will likely be myriad issues to consider/address relating to the advent of advanced nanomedical technologies with the potential to dramatically enhance the full range of human cognitive capacities. This sophisticated interface might serve as a bespoke portal through which individuals could not only have instantaneous access to every aspect of collective human knowledge but also have options to engage in real-time, fully immersive experiences. Further, a B/CI may enable opportunities to episodically 'inhabit' any other consenting individual on the planet in real-time, fully sensorial resolution, via what we refer to as 'transparent shadowing' (TS). This chapter also explores how a B/CI might impact one's sense of self and how that self might relate to others, the world, and beyond. The potentially ubiquitous implementation of a B/CI may be nothing less than the next evolutionary step for humanity, one where we seamlessly merge with our technologies.

1.1 Introduction

The notion of a hypothetical (for now) human B/CI posits that it might be technologically and nanomedically feasible within the next few decades to seamlessly interface the human neocortex with multitudinous highly distributed (cloud/edge) supercomputers. A direct B/CI enabled by perhaps billions of autonomous self-migrating and self-installing neuralnanorobots or a less intrusive wireless, cranially embedded, nanorobot-mediated B/CI might facilitate a considerable range of applications (Martins *et al* 2019). Users would have the capacity to instantaneously

doi:10.1088/978-0-7503-2144-0ch1

access any digitized textual or graphic information, recorded or live sound and/or video, as well as the discretionary capacity to engage in real-time, fully immersive virtual travel, films, conferences, performances, and more.

One of the most intriguing potential applications of a B/CI might involve what we refer to as TS, whereby, under strict security and procedural protocols, voluntary or remunerated certified TS 'spatial hosts' (SHs) would be available for single or multiple attendees (ATs) (conceivably numbering in the millions) to literally experience episodic segments of their lives on a predetermined time schedule. These TS sessions might take the form of seminars, courses, or lecture series, where the insightful knowledge or specific skills of the SH may be imparted to the ATs. Further, they may include seamless experiential resolution that encompasses the full sensorial realm, which would be experienced by the ATs as if they inhabited the physical body of the SH. Although the ATs would perceive the vocal instructions of the SH, as well as sensually and temporally experience exactly what the SH is experiencing, they would be completely blocked from any access to the SH's thoughts, emotions, or self-speak. From the perspective of the SH, the only indicator that he/she is 'live' might take the form of some type of visually overlaid indicator icon.

1.2 Potential ethical considerations for a brain/cloud interface

1.2.1 Preamble

The established field of nanoethics (Biroudian *et al* 2019, Sunshine and Paller 2019) has valiantly attempted to provide cogent guidance toward the development of unknown future technologies. There are a multiplicity of challenges associated with the preemptive formulation of strategies, validation of perceived rationales, and responsible preparation for situations that may not emerge downstream (in some cases) for decades. Nevertheless, these well-intentioned efforts proceed, even when their participants fully acknowledge that the reality of the future they are planning for and attempting to safeguard against may indeed be quite disparate from what is envisaged from afar. Distortions can certainly arise, akin to those that appear when we seek to identify the detailed features of distant landscapes. This should be considered in conjunction with the potential futility of attempting to prepare for unknowable futures when myriad interim events can singly or cumulatively alter downstream trajectories, sometimes radically so. Useful ethical formulations, regulatory intelligence, and the long arc of societal maturity often seem to lag (sometimes woefully so) behind technical progress. While acknowledging these challenges and the fact that there is already a significant body of work in the nanoethics domain (even if it is a false umbrella term) (Grunwald 2010), this chapter presents speculative explorations and views that may not necessarily align with existing channels of nanoethical investigation.

1.2.2 Ethics of brain-embedded medical neuralnanorobots

The self-emplacement of perhaps billions of micron-scale autonomous neuralnanor-obots within the human brain to create a B/CI presents ethical issues that parallel those raised in conventional medical treatments and surgical environments.

The decision-making capacities of potential B/CI users will clearly be a critical consideration, as obtaining informed consent for the procedure will be mandatory. Typical guidelines employed in clinical medicine may be instructive in showing us how to approach potential ethical issues with B/CI technologies (Appelbaum 2007). Notably, the strong potential for B/CI technologies to positively impact the quality of life of those burdened with disabilities creates additional ethical issues akin to current concerns. Importantly, careful considerations must be observed prior to B/CI being offered as a 'cure' for some disabilities, as not all persons with disabilities wish to be cured (Harman 2023). For some individuals with certain disabilities, their disability is an integral part of their identity; thus, it is a difference to be accepted, not a disease to be cured. A potential example would involve autism spectrum disorders and the associated neurodiversity movement, which rejects the search for a cure (Jaarsma and Welin 2012). It is unclear how the autism and neurodiversity communities would respond to the widespread use of B/CI technologies to alter the functioning of individuals with autism; however, ongoing research in this field is promising (Friedrich *et al* 2014).

1.2.3 Ethics of transparent shadowing

A unique application of a B/CI, referred to as TS, would involve B/CI users engaging with accredited 'spatial hosts' (SHs), who could opt to volunteer or be remunerated. A series of ethical questions may be raised regarding the protocols and meticulousness of SH screening and selection, accreditation, and monetary exchange. The process of determining the eligibility of SHs must take into consideration multiple factors to ensure fair and equal access across the full range of anthropological groups. One issue of serious concern that must be addressed, particularly in view of current events, is that those voices that have historically been suppressed or silenced, as well as typically marginalized cultural groups, may have their experiences further excluded from public view/experience.

The selection of SHs should accurately reflect and accommodate the extensive range of cultural diversity, perspectives, and experiences found across the globe. The SH accreditation process should strive to maintain this diversity as well. The issue of potential payment for services as a SH might present a number of ethical quandaries (e.g. impoverished SHs selling their experiences of poverty to wealthy As, possibly under agreements with charitable agencies, as part of a new form of 'poverty porn'). Careful consideration of these and other ethical concerns to support the development of stringent, yet inclusive and fair, protocols for the SH selection process is strongly recommended.

1.2.4 Warning against an overly cautious ethics

One position espoused by bioethicist Allen Buchanan was a worry that beneficial technology development for cognitive enhancement, such as brain–machine interfaces (BMIs), remains mired in an overly cautious ethics that is too slow and too resistant to change (Buchanan 2011, 2013). Instead, Buchanan positively framed the 'enhancement enterprise' to counter the typical inhibitory debate against enhancement. He noted that society gives freedom to individuals and groups to select, develop, and implement these types of enhancements and calls for the allocation of resources toward the development of a public that is well-informed on these issues. Enhancement is not 'unnatural'; it is perhaps the most natural of all enterprises, as humanity has always sought to improve/advance itself. Nature is not some delicate construct that is best left undisturbed; indeed, it is in a constant state of flux, interruption, and emergence. Some of the most significant life processes, such as natural selection and reproductive fitness, are not themselves targets of cognitive enhancement.

1.2.5 Working toward an ethics of immanence

It is easy to see why new technologies that approximate BMIs and cognitive enhancement have been errantly perceived as being (within the philosophical frame) exclusively an issue of bioethics. Ethics, as generally conceived, is undoubled, spatialized, tangible, visible, easily articulable, research-fundable, and policy-translatable. This traditional undoubled ethics, as a moral philosophy, is valorized by the apparatus of control society, against which modern efforts at self-responsibility struggle, as exemplified by Brin's *Transparent Society* (Brin 1998), Foucault's *Panopticon* (*Discipline and Punish*) (Foucault 1977), and Deleuze and Guattari's *Anti-Oedipus* (Deleuze and Guattari 1972). An immanence ethics completely restores the side of reality that resides on the other side of the baseline and instantiates the unlimited upside of human potential and actualization (Swan 2015a).

1.3 Brain/cloud interface—neural trust, blockchain, and neural cryptographic security

1.3.1 Neural trust

Neural dust must come with neural trust.

—Jan M Rabaey.

Trust will be a central issue for the successful implementation of a B/CI. In relation to cybersecurity, personal data, and manipulative neuromarketing concerns, reliance on third parties (or any external model of trust) will be a non-starter. Science fiction has already painted this picture to avoid in terms of the disruptive effects of neural hacking (Bear 1997). An advanced and mature B/CI must include onboard functionality for the generation of trust and cybersecurity protection. This section discusses how neural trust might be a resource that is provisioned by B/CI operating software. The argument is that B/CI platforms, as with any advanced computing system, are likely to have built-in features for cryptographic data protection as part of their standard operation.

Neural prostheses will become completely integrated with our perceptual apparatus, such that we will notice immediately if something is amiss. Merleau-Ponty (1945) considers examples of eyeglasses and the blind man's stick. In relation to

B/CIs, failsafe strategies will be essential to earn the trust of users in their installed augmentation. An analogy is scuba diving, for which we learn how the life-support equipment operates, we train with it, develop trust in it as our prosthesis, and work with it in situ such that (hopefully) issues can be immediately diagnosed and resolved. The same will be required for B/CIs.

The B/CI consists of a dedicated computational platform integrating software protocols, standards, and security. For cloud-connected B/CIs, numerous details regarding the data transfer rate, protocols, communications software, and security will need to be established. Thus far, three basic levels of B/CI applications have been outlined. These include the existing functionality of controlling prosthetic limbs with electrical signals from the brain, the near-future interfacing of brains with the internet at broadband speeds to interact with external software and devices, and the longer-term notion of a full-blown B/CI that includes brain-embedded (or brain-peripheral) medical nanorobots. In the existing applications, some B/CIs or neuro-prosthetics have the capacity to interpret brain signals (electrical pulses) to enable individuals with disabilities to control prosthetic arms and legs, potentially with some level of internet connection. For example, Hanger, a Texas-based provider of orthotic aids and prosthetic limbs, collects near real-time data regarding usage and mobility, connecting directly to AT&T's 4G Long-Term Evolution for Machines (LTE-M) network (Scroxton 2018).

It is estimated that full-fledged B/CIs will require 'broadband access with extremely high upload and download speeds compared to today's rates (Martins *et al* 2019). Internet networks are already beginning to accommodate two-way transfer. Although initially designed for asymmetric data downloads from servers to clients, communications networks now support large volumes of data being uploaded from Internet of Things (IoT) sensors and consumer devices. The projected data rate for a prospective B/CI-class upload would be 24 Mb s^{-1}, which would take place via both Bluetooth 4.0 (for the IoT) and IEEE 802.11n low-power Wi-Fi technology (for body-area networks (BANs)) (Zao *et al* 2014). Next-generation communications networks such as 5G (100–200 MB download speeds) and anticipated terahertz networks (100 GB data links) may play a role in the ultra-high-speed wireless data networks of the future, which could accommodate the required upload and download speeds for B/CIs.

As network transmission rates accelerate, data protection is beginning to become a standard feature of internet networks, as mandated by General Data Protection Regulation (GDPR) compliance in Europe (legislated in 2016) and other initiatives. The market research firm Digital Universe indicated that, as of 2012, less than 20% of the world's data was being protected (EMC Corporation 2012). Privacy protection, including the use of cryptology-based strategies such as blockchains and zero-knowledge proofs (proving access to a secret or key to transmit or access data), might become a network computing standard. A completely new level of data protection could be a universal feature of any computing system by the time more mature and robust B/CIs are implemented, perhaps in ∼20–30 years (which may be significantly expedited in view of the recent rapid advance in AI) (Martins *et al* 2019). Such concerns have been framed in the current design standards of B/CIs for medical applications, which call for signal-acquisition hardware and software that is convenient, portable, safe, and able to function in all environments (Shih *et al* 2012).

1.3.2 Blockchain-based neural trust

One potential neural security solution for advanced B/CIs is proposed here, which specifically elaborates on how a blockchain system might be employed to create trust in a B/CI software network. A blockchain is a secure ledger system in which transactions are confirmed automatically by software within a flat peer-to-peer network architecture (Swan 2015b). Every entity (e.g. neural nanorobot) in the system would be assigned an address (a unique identification number) to ensure that all its activities are automatically approved and logged as transaction records by the network. Since any activity must be confirmed by the network prior to execution, this would be the source of trust, referred to as 'algorithmic trust.' Transaction confirmations would occur through dedicated peer nodes in the network that oversee the confirmation of transactions, which would be distinct from the nodes that undertake given transactions. Tasking confirmations would be automatically processed by the network software (via the consensus algorithm) only if the conditions for the transaction were met. Valid transactions would be confirmed, executed, and batched together in blocks (hence the name blockchain). Security and trust are provided by virtue of the fact that the network is a highly localized system (i.e. in-brain) where any transaction must be confirmed *in situ* before it is executed, having been submitted by valid system entities with the requisite permissions and resources for the transaction.

Blockchains operate in the macro world, with the network software automatically confirming and executing transactions. The largest blockchain to date, Bitcoin, has been operating for over 16 years (since January 2009) and, as of July 14, 2025, has confirmed over 905 531 blocks of transactions (BitInfoCharts 2025). Thus far, the primary focus of blockchain transactions has been monetary transfer (carried out instantaneously on a global basis), so the question is: what type of currency would be transferred within B/CI networks? Consider that blockchain operation requires each entity in the system to have an address (a unique identifier akin to an email address). Any address can have different currency balances associated with it for its activities. For example, one address (or 'wallet') might have a U.S. dollar balance for USD activities, a Euro balance, a Bitcoin balance, etc. The blockchain ledger system is simply an immense online global database that stores information (in a protected way that can only be transacted in certain ways). To give the most basic example, the blockchain ledger is an online record of who owns what (e.g. Tom owns $5, Sally owns $7). However, the ledger cell is a database record that can store multiple different types of 'currencies,' such as an account balance, an auto title, digital rights to music, Netflix subscription details, medical test records, academic diploma credentials, or computer code that can be called by other programs.

In a B/CI network, the relevant currency ('wallet balance') associated with each neural nanorobot address is the mechanism that enables the neuralnanorobots to perform their permitted activities. For example, neuralnanorobots may need

Table 1.1. Classes of human B/CI traffic.

Application class	Application	Functionality	Traffic type	Ledger currency
Current applications	Neuroprosthetics	Actuation: limbs	EEG signals	Microvolts
Near-term applications	Internet access	Information search	HTTP requests	MB
	Data upload, backup	Data privacy	HTTP POST/ GET	MB, service level agreements (SLAs)
Enhancement: long-term applications	Brain-embedded medical nanorobots	Cybersecurity, fleet management	Blockchain, zero-knowledge proofs	Millivolts, millimolecules
	TS	Sensory experience-casting	Quantified self biosensors	Millimolecules, experience units
	Cloudmind	Collaboration and credit-assignation	Intellectual property (IP) logging	IP units, ideas, novelty

electricity currency balances (to generate signals and potentials), neurotransmitters, and internet traffic (to POST their status). The B/CI network will need to accommodate different layers of traffic (table 1.1). At a minimum, the different types of traffic flowing through the B/CI network might include internet traffic, electrical signal traffic, and neurotransmitter traffic. Just as the internet transfers different layers of traffic using different software protocols (e.g. data, voice, and video traffic), so too will the B/CI network need to transfer different types of traffic that are relevant to its activities. Each traffic type will have its own software transfer protocol (i.e. operating instructions), currency, and measurement metric. Electricity is one neural currency, denominated in microvolts, that allows the transfer of signals and the provisioning of field potentials. Another neural currency is neurotransmitters (e.g. serotonin, dopamine, and gamma-aminobutyric acid (GABA)), which are measured in millivolts. Internet traffic is another currency, measured in megabytes (MB), or postconversion electrical signals that bypass (costly) rod and cone visual processing (optogenetic therapies provide a model for bypassing retinal processing (Williams 2017)).

The network nodes, or neuralnanorobots, serve as the agents that transact the B/CI traffic as per preprogrammed instructions, doing so in an approved and auditable way by having a currency balance as the mechanism by which each node conducts its activities. Thus, economic concepts are the design principles for the B/CI network operating software. An analogy is that of giving each student $10 of lunch money toward achieving the collective goal that the class has lunch. Likewise, serotonin balances are distributed to neuralnanorobots, which coordinate their activities toward the group goal of reducing depression. The practical application is using smart contracts (blockchain-based code) to activate neuro-transmitters on a more granular basis to manage depression without the side effects of prescription drugs.

A typical B/CI design task might be to identify a list of the different brain regions and specialty fleets of nanorobots that will target them for specific applications, along with the types of requirements these applications will have (e.g. network security and data transfer rates). Every traffic class will have its own currency stored in wallet balances on system nodes, for example, to access internet data in a data lookup. Presumably, only certain perceptual regions of the brain and classes of B/CI nanorobots will be involved in major internet data search and retrieval tasks. Certain classes of neuralnanorobots may be deployed in specific brain regions prior to the self-installation of standardized B/CI 'packages' with the aim of optimizing nominal brain function. They might be involved in particular activities, such as the rebuilding of memories from off-site storage to support stroke rehabilitation, or biomedical operations such as amyloid plaque or lipofuscin degradation and removal. There may well be a list of B/CI neural nanorobot classes and brain operations that are categorized by feasibility and risk, with roadmaps for imple-mentation (table 1.1). Initial low-risk applications could include neural mapping, health scans, and memory backup. Notably, the same B/CI blockchain network operating software with transaction-logging functionalities might be utilized for cryptographic security, as well as in support of later-stage IP-tracking applications to assign credits for cloudmind collaborations.

1.3.3 Billion-neural-nanorobot orchestration and system control

Once instructions are specified (via programmatic smart contracts), a B/CI block-chain network would simply run the instructions (as a Turing machine). Security (and trust) would be created in that the network would only have the capacity to execute its preset instructions, and nothing more. The benefit of blockchain systems is that they are both operationally necessary and security-providing. Some type of global coordination mechanism would be required to control likely billions of neuralnanorobots. Blockchain is precisely such an automated operating system that could arbitrarily accommodate multiple nanorobotic fleets. A further inherent feature of blockchain systems is that they provide security, as only transactions that are locally confirmed by the system may be executed.

Looking further ahead, improved control capabilities for managing billions of many-particle fleets of brain-embedded medical nanorobots may be available from physics. The implication of the billion fleet-many nanorobots being logged in a blockchain system for coordination is that they can be controlled in an automated manner, at various higher levels of abstraction that correspond to system states. Physics has mathematical models that are readily equipped to manage scenarios involving many-particle systems; for example, by providing one metric at a higher level of abstraction that can be employed for system control. Room temperature is such a macro-level value that corresponds to the overall state of the micro-level activities of all the particles in a room. Likewise, the blockchain 'temperature' or 'Hamiltonian' (multidimensional composite value of the state of a system) could be derived as a control parameter. This could be useful as a mechanism for managing systemic risk, i.e. as an immediate shut-off controller for all nanorobot activities.

1.3.4 Questions, risks, and limitations

Several concerns regarding B/CI blockchain networks relate to speed, scalability, and efficacy. First, regarding speed, would there be sufficient time for a peer-to-peer network to confirm microtransactions for even a few millimoles of neurotransmitter that needed to be deployed immediately? The answer is that the transaction confirmation can happen as quickly as the instructions are sent, as part of the instructions. B/CIs would essentially solve a signal processing problem, and the control mechanism is that the signal is not sent unless it has been confirmed. Security is just another step in the software operating process. Furthermore, certain classes of low-risk transactions, analogous to petty cash transactions in a business, might be preapproved (such as sustaining minimal levels of a neurotransmitter or maintaining a field potential). This could include the use of blockchain features such as payment channels (streamlined, automatically settled transfer mechanisms offered by projects such as the Lightning Network) (Martinazzi and Flori 2020).

Second, regarding scalability and efficacy, a concern might arise as to whether a B/CI blockchain system can be efficient in terms of consensus and storage. Efficient consensus is already an issue in macroscale blockchains, where the core consensus method in operation is the proof-of-work algorithm, which entails a competition between nodes to win the right to confirm the latest transaction block for an economic reward. Competing nodes prove their intent is non-malicious by performing computational work (proof of work). Although proof-of-work consensus succeeds in providing cryptographic security, it is expensive; thus, perhaps not scalable. Therefore, B/CI blockchain systems might implement next-generation consensus algorithms that are far more efficient.

There are already different species of consensus algorithms for different classes of blockchains (e.g. money transfer chains (Bitcoin, Ethereum) use proof-of-work mining, and IoT chains (Iota, Hashgraph) use other protocols). IoT consensus algorithms may be suitable for use as part of the B/CI blockchain network protocol to provide the three classes/layers of required neural cryptographic protection: in-brain security, brain–cloud security, and brain–other mind security. The security is provided by virtue of the fact that no transaction can be processed on the network without having been confirmed by other nodes that are unrelated to the transaction.

For mining (transaction confirmation and security), B/CI blockchain networks might have nanorobot nodes assigned for transaction confirmations only or have peer-based ecologies that mine for each other (nanorobots in one brain region confirm transactions for another). For example, visual perception nanorobot nodes could mine for those of hearing. There will likely be different classes of nanorobots deployed to various areas of the brain for different purposes. The economic design principles of macro blockchains can be reused in the B/CI blockchain environment.

In B/CI blockchains, peer nanorobot nodes mine (confirm) transactions for other nodes for a small fee. Transaction confirmation is a 'utility function' of security and accountancy, performed by the peer network. It is not free, and therefore a small cost-based fee could be paid to the confirming nodes by the transaction-submitting nodes (a 1% transaction fee is the standard). This raises the complicated question of exchange between neural currencies, which is a standard feature in blockchain systems.

However, rather than attempting to manage a conversion from millivolts of electricity to millimoles of neurotransmitter, perhaps all the electrically based nanorobot nodes could mine within one ecological subsystem. A proof-of-stake (instead of proof-of-work) neural mining consensus algorithm might make more efficient use of resources. In the proof-of-stake model, participating mining nodes stake a large currency balance for the duration of the confirmation round to prove to the network they are bona fide actors (proof-of-stake), which is recovered when the confirmation round is complete. The requirement is to have an automated software-based apparatus in an open network environment that creates a system in which good actors can prove their intentions and bad actors cannot operate. Macro-level blockchains and B/CI blockchains might deploy the same type of structural mechanisms to prevent bad player behavior and thus provide security and trust.

There remains a question about where the bioenergy is to come from in the mining operation (transaction confirmation). However, this would be part of the overall energy cost of the B/CI network operating software and priced into the cost of the network utility function used to provide confirmation. The underlying question is whether the brain has the energy to spare and if the value proposition of spending a little bit of energy for neural cryptosecurity is worth it, where clearly the answer is yes. The cost of B/CI network operating software (including security) must be priced into the overall cost and in situ energy budget of neural nanorobot deployment.

Another efficiency concern is storage, as logging every change to a nanorobot's neurotransmitter levels could quickly consume a lot of storage. The B/CI network might include billions of nanorobots, each with millions of daily transactions. The solution is to learn what data needs to be retained and for how long. Blockchains are, by definition, a permanent immutable ledger, albeit transaction details could be off-loaded to sidechains (as with signatures) or logged to decentralized storage (such as InterPlanetary File System (IPFS)) that could be purged and reused. The point is that all transactions can be controlled in terms of their execution in the first place (which confers security and trust) and can be logged for record-keeping and audit purposes. The automated record-keeping functionality exists; thus, it is merely a decision about how the functionality is to be managed.

Certainly, there are other high-stakes risks that are difficult to evaluate at present but will be essential to address as a prerequisite to the widespread implementation of a B/CI. One such risk is that of electromagnetic pulses (EMPs) and live, mind-based internet connections. This is a use case for the blockchain Hamiltonian that serves as a global B/CI system override or off-switch, essentially as a neural surge protector. Software updates represent an issue that needs to be clarified: exactly who can update the software and how? The analogy in the macro blockchain world is the democratic method of open-source developer communities and a structured process for submitting and voting on software improvement proposals that are then implemented by the worldwide community. Intrusion detection is another issue: how do you know that a smart contract virus has not issued itself a network address and infiltrated your B/CI system? How can a malicious smart contract obtain a network address? These and other issues will need to be resolved. New technologies typically evolve in lockstep with threats that work against them. The good news is that the early-stage learnings from using cryptographic networks for monetary transfer may translate nicely to other, even more sensitive domains, such as the human brain.

Given that trust will be of critical concern for a potentially ubiquitous B/CI implementation, a specific solution is outlined here as a B/CI blockchain network, which would provide 'neural trust for neural dust' (B/CI neuralnanorobots). The argument is that the advanced computing networks of the future are likely to have robust and dynamic quantum cryptographic data security standards that today's networks do not. As with any advanced computing network (~10–30 years from now), software that is being considered for B/CI operations will likely be intrinsically replete with standardized advanced (perhaps nested) security features. Blockchains are one form of advanced cryptographic network software that can provide both cybersecurity and the automated control of perhaps billions of autonomous entities (neuralnanorobots) that will be accessing our brains and operating within them.

1.4 Brain/cloud interface—potential moral considerations

1.4.1 Morality of brain-embedded medical nanorobots

1.4.1.1 Agency and identity
An *agent* may be considered an autonomous actor that persists over time, who can be held responsible (causally, and potentially legally and morally) for his/her/its actions (Roskies 2015). Since humans are moral agents, the autonomy of individuals could potentially be violated by future B/CI systems. In view of the recent (2023) dramatic advances in AI via ChatGPT, etc, in conjunction with the rapid improvements in quantum computers and the potential relatively near-term emergence of artificial general intelligence (AGI), computers may be imbued with intelligence that is equivalent to, or far surpasses, that of humans, which may have impacts on human autonomy when individuals are cognitively engaged via a B/CI. In light of the above, it might be speculated that the core AI-mediated B/CI infrastructure itself could at some point become conscious and self-aware, which would undoubtedly give rise to myriad serious ethical, moral, and even human existential issues. How might we ensure that this consciousness emerges as (and steadfastly remains) benevolent while operating on the premise of benefiting humans and humanity?

Unexpected effects of a BCI on agency and identity may also occur. Gilbert *et al* (2019) reported on six patients with medically intractable epilepsy who were implanted with the first in-human 'BCI advisory devices' for seizure prediction. Although BCIs can positively increase a sense of self, confidence, and control, one patient suffered from distress, feelings of loss of control, and a 'rupture' of patient identity or self-estrangement. In the future, this may be less of an issue with the augmentation of normal functions via B/CI devices, in contrast to applications for neurological disability. A standardized psychological preparation of potential patients to build resilience and prepare them for the likely effects on self-image should be developed and implemented for future B/CI recipients.

1.4.1.2 Neural privacy

Issues of agency related to a B/CI may pertain to the privacy of neural data. For instance, who would own the thoughts generated in these networks, particularly when they might lead to valorized/monetized innovations, new products or services, etc? Furthermore, who or what agency would be responsible for flagging, decoding, and registering these actions, and how might this cognitive output be authenticated? Who is responsible for actions? What is the authenticity of the output? What will AI contribute to these inputs/outputs? Agency and identity are human rights that must be protected; therefore, discussions of agency and identity should be included in the consent process for the implantation of neuroprosthetics. With the advent of a hypothetical (for now) nanomedically enabled human B/CI, any capacity for 'mind reading' should be soundly rejected on ethical/moral grounds.

1.4.1.3 Stigma and normality

Although the aims of a B/CI intended to address cognitive disabilities will be to improve neurological functionality and quality of life, the actual situation might be more complex and nuanced. Will disabled individuals decide to engage a B/CI solely to reduce disease stigma or to reduce their perceived burden on society? Conversely, some users may feel increased stigma when using a B/CI and opt to restrict its use. Furthermore, some disabled people simply do not perceive themselves as disabled; thus, B/CI would serve as an enhancement to them rather than a treatment. Beyond its use to treat various neurological conditions, what further issues may arise when a B/CI is employed for the augmentation of normal cognitive function? Individuals who demonstrate perceived superhuman qualities via a B/CI would readily attract negative attention, envy, and even physical abuse. Additionally, these supersoldiers or superintelligence agents would be considered high-value targets by adversaries.

1.4.1.4 Brain/cloud interface user autonomy

Autonomy may be significantly undermined by severe neurological disabilities. On one hand, a B/CI could empower an individual by giving them more independence, thus improving their autonomy and dignity. Conversely, when a B/CI plays causal roles in the decision-making processes of individuals, this may further affect autonomy, which may be even more of an issue when considering the massive/simultaneous global-scale B/CI data transfer capacities involved. It is also conceivable, even if extremely unlikely, that an external agent might somehow manage to exert mind control over an individual by hacking into a B/CI through quantum-encrypted security systems, a theme extensively explored in various science fiction novels and movies.

1.4.1.5 Brain/cloud interface user responsibility

A further uncertainty relates to whether B/CI users will be solely and entirely responsible for their cognitive output once engaged with/immersed within these systems. Since legal edicts have not kept anywhere near pace with the advances in today's BCI systems, the attribution of ultimate responsibility for the events that occur in exponentially more powerful B/CI systems is unclear. Where do the legal and moral responsibilities for the behavioral effects of B/CI users lie? One view is that these individuals should be held fully liable for any unintended consequences. For example, the passing thoughts of a given user might trigger a cascade of unwanted or even damaging responses. This raises the question of whether certain cognitive filters (e.g. such as the time delays employed for live public events) should be incorporated into B/CI systems as a standard feature to negate the possibility of these issues ever occurring. Others may posit that these undesired effects might be due to the B/CI system itself, in which case the manufacturer would bear some responsibility. This will be a complex interrelationship, contingent on the given situation. The potential, however rare, for a third party to hack into the controls of B/CI systems would add further complexity.

What if the robotic arm connected to an individual's brain by BCI injures a bystander? Was it the unconscious thought of the individual that triggered the action, or was it their conscious malicious thought, or was it the semi-independent robot? The BCI user should be able to cancel the action of the robot using a 'veto switch' activated by eye movements. Clausen *et al* (2017) suggest that all new BCI systems should include a veto control. Inappropriate use of the device by the individual may make them liable for causing harm. The B/CI medical team and the manufacturer also have a responsibility to train the user in the correct and safe use of the device.

1.4.1.6 Informed consent

The invasiveness of conventional BMI and BCI neuroprostheses raises safety and ethical issues. Even the most advanced brain electrodes to date, proposed by Neuralink (Musk 2019), are quite invasive, as they require the permanent removal of a portion of the skull approximately 25 mm in diameter and the emplacement of a skull 'plug' along with multiple ultrathin electrode threads that are surgically implanted via a robotic 'sewing machine.'

It is envisaged that the three species of neural nanorobotic devices (endoneurobots, gliabots, and synaptobots) might be introduced into the human body through various means, which may include injection, intranasal spray, aerosolization, oral administration as a pill or syrup, or via a dermal patch or topical gel. These neuralnanorobots will be preprogrammed to self-migrate to their assigned positions as determined from previously acquired ultrahigh-resolution imagery of the brains of individual patients or healthy users. Since these nanorobots will be so diminutive ($\sim$1 μm in diameter), their entry into and migration through the cerebral cortex will be physiologically imperceptible to the patient/user. Thus, they may be considered essentially physiologically noninvasive.

A conjunctive wireless signal reception/transmission mesh would also be comprised of self-migrating nanorobots that shallowly self-position themselves within the skull or directly above the skull under the scalp dermal tissues. Although the self-installation of a B/CI would be a relatively major medical undertaking, which might require an hour or several hours to complete, there must be an acceptable risk/benefit ratio. The complete procedure must be fully disclosed and clearly explained to the potential recipients and their families beforehand, with signed consent given. Informed consent should include a comprehensive discussion of the possible effects of a B/CI on mood, personality, or sense of self (particularly concerning any residual influences of TS) (Yuste *et al* 2017).

1.4.1.7 Managing expectations

Regarding the self-installation of B/CI systems, a realistic explanation should be provided of what is involved in any potential risks associated with preliminary brain mapping, pre-installation screening, neural nanorobotic self-migration procedures, post-installation testing/verification protocols, and the procedure for full uninstallation (if desired downstream). Such an explanation should be essential. Researchers, manufacturers, and regulatory agencies will have the responsibility to accurately and transparently communicate the procedures, features, and operational details of B/CI platforms. It is likely that an inbuilt 'B/CI owner's manual' will be standard programming for the neuralnanorobots, which users can summon at any juncture during any application by simply making a mental query.

Informed consent should include a discussion of the possible effects of the device on mood, personality, or sense of self (Yuste *et al* 2017). Therapeutic misconception is a belief by the patient that they will receive benefit from a given intervention. This may not be the case with a B/CI, which may not always be effective. The device may fail to deliver; thus, the user needs to be prepared for the failure of the B/CI to achieve anticipated goals. Any hyperbole or exaggeration regarding the operation and applications of a B/CI will be unethical. For example, media/advertiser portrayals of B/CIs might tend to be overly zealous and optimistic. All aspects of B/CI protocols and operations should be transparent, and any misinformation should be quickly identified and eradicated.

1.4.1.8 Brain/cloud interface beta testing: balancing risks and benefits

Once a neural nanorobotically enabled B/CI platform has been realized, extensive discussions of the risks versus benefits with the participants of beta-phase testing (and their families, if applicable) will be mandatory, as this will be an unprecedented and particularly powerful technological platform. There should be full disclosure of all available data, and the volunteer/patient/user should have the capacity to clearly understand the information presented in a deliberate and appropriate manner. Furthermore, B/CI users should be free from any coercion or undue influence in making the decision to proceed. Participants would reserve the right to withdraw from beta testing at any time, preferably prior to autoinstallation.

It is important that in these discussions, the physician provides the patient with hope, but it should be based on 'realism' rather than 'idealism' and placed in perspective. The psychological and physical condition of the patient must be considered, while psychologists, psychiatrists, and physicians should be involved in the screening of potential recipients. Cognitive or affective impairment would be a contraindication for an implantable B/CI, except if in the future it is used specifically to boost cognitive function, such as a memory prosthesis. B/CIs may become less reliable in the presence of cognitive impairments, or they may result in aberrant communication with the device. Researchers who are implanting a new device into a patient have a duty of care to that individual for the long term. Further, the neurosurgeon should be prepared to manage complications whenever they might occur and be prepared to deactivate the B/CI or initiate its auto-egress if requested by the recipient.

1.4.1.8.1 User safety

Since the physical dimensions of the various species of autonomous neuralnanorobots that cumulatively comprise a B/CI would be on the order of ∼1 μm in diameter, their self-installation or self-uninstallation will be physiologically imperceptible to users. At this size range, the nanorobots would avoid any contact with the nervous system, save for when they interface with the neurons of the brain. Neuralnanorobots would likely be designed to be essentially spherical in shape and morphologically smooth, thus negating the possibility of any physical damage to cells, vasculature, and tissues. Furthermore, they would comprise biocompatible diamondoid or sapphire materials to negate the elicitation of an immune response.

In the regulatory realm, stringent independent assessments and quality control measures associated with the design, engineering, fabrication, computational, and software elements of B/CIs will be required prior to their use for human applications. The nanomaterials employed for B/CIs will need to be International Organization for Standardization (ISO) certified and approved by the appropriate regulators. Furthermore, permissions to perform the self-implantation of B/CI neuroprostheses must be cleared by the U.S. Food and Drug Administration (FDA), be legally and jurisdictionally sanctioned, and be approved by the human research ethics committees of the various global medical institutions for the regions where the procedures are to take place. Qualified neurosurgeons/clinicians will also need to be approved to perform these procedures by medical institutions.

The short-term risks of implantation, which may include infection, hemorrhage, and epilepsy, must be fully explained. The likely functional longevity of the prosthetic must also be explained. The long-term risks of the implanted device include device failure due to gliosis around the electrodes or loss of hermetic sealing, exposing the internal electronics to fluid damage and the potential leakage of any toxic materials such as heavy metals from the device (note: it is likely that advanced autonomous nanomedical devices will contain no toxic materials, as the computational and electronic elements might be photonic and carbon-based, respectively) into the circulation around the electronics must also be explained. Epilepsy and infection are also longer-term risks. It would also be useful to have patient and user registries established, such that all implanted BCI neuroprostheses can be tracked over time; such registries could be blockchain based.

1.4.1.8.2 Inappropriate expectations

The potential benefits and drawbacks of a B/CI will need to be put into perspective and clearly explained in terms of the likely quantum improvements in neurological functions and activities of daily living, should the system operate as expected. Patients/ users will also need to be made aware of the possibility (albeit remote) that the system might not operate properly at times or may completely fail due to unforeseen circumstances. Participants in the first B/CI human trials should not be expected to pay fees for any aspect of the study or device. On the contrary, they should be remunerated by the companies/research facilities that are offering the platform. Further, any alternative assistive devices should be explained in detail to the participants. Lane *et al* (2012) surveyed potential recipients of bionic vision devices and reported that they may perceive themselves as pioneers, trailblazers, adventurers, and explorers; thus, they should be welcomed as integral collaborators in the research, as there are likely to be very small numbers of participants in the first in-human B/CI trials.

B/CI neuroprostheses are likely to be superseded by newer technologies in short order. For the most part, this will be the case with the emergence and broad implementation of AGI and artificial superintelligence (ASI). In the case of a newly developed neuroprosthesis that is being initially implanted by craniotomy as a one-off procedure, the investigators would need to explain to the recipient what benefits they are likely to derive, but should also explain that they are pioneers and will be assisting investigators to achieve a better understanding of the device functionalities and capacities, even if it may be superseded in the future. Rehabilitation for B/CI users may be arduous and extensive and include scientific research conducted by the investigators. This may place physical, emotional, and financial burdens on the individuals and their families, which need to be explained to the potential recipient. A slow pace of expected results may result in frustration, lack of motivation on the part of the patient to cooperate with the investigators, and possibly depression. Failure of the device could have a serious emotional effect on the user, particularly if they had been expecting to depend on the technology. Currently, approximately 15%–30% of users with disabilities fail to use a BCI effectively, which has been referred to as 'BCI illiteracy' (Klein 2016). This failure rate may be substantially lower when applying B/CIs for augmentation.

Independent assessments of engineering and manufacturing quality control are necessary prior to human medical applications, while the materials used for implantation need to be ISO certified and approved by regulators. Permission to perform the implantation of neuroprostheses is typically obtained from the institution's human research ethics committee. The hospital involved will also need to accredit the neurosurgeons who are to perform the procedure.

Patients who seek to engage with a B/CI specifically and exclusively for the treatment and/or monitoring of certain cognitive conditions (e.g. Alzheimer's disease, Parkinson's disease, epilepsy, etc.) should be thoroughly briefed on what to expect in terms of realistic clinical outcomes and associated timelines. Consequently, this might place certain physical, emotional, and financial burdens on the patient and their families, which would need to be comprehensively explained. (*Note: these burdens might be alleviated with the advent of an envisaged equitable, cost-effective, non-monopolizable, globally distributed AI/QC-driven healthcare system (global health care equivalency (GHCE)) operating on nested blockchains*).

1.4.1.8.3 *Placebo effect and the establishment of efficacy*

The placebo effect might be an issue for patients with disabilities who undergo B/CI installation; however, it might also conceivably occur for augmentation applications. Placebo effects might occur following the installation of a B/CI, which could inflate the outcomes of their use. Increased support from the therapist/physician of a disabled patient following the installation of a B/CI neuroprosthesis will likely have positive effects on patient performance. However, the patient might enhance their own recovery to some degree when the B/CI is disengaged.

1.4.2 Morality of transparent shadowing

Transport shadowing (TS) (a B/CI application through which we may have the capacity to episodically 'inhabit' any other consenting/certified B/CI user on the planet in real-time, with full sensorial resolution) raises some intriguing questions involving morality. For example, the potential for crime could result in moral and associated legal predicaments. Consider the following scenarios. If an SH has the misfortune of becoming the victim of a violent crime during a TS session, how should the AT or ATs respond? Will there be a moral obligation to attempt to determine the exact location of the SH, correctly identify the SH, and immediately contact the proper authorities? Should this contingency be integrated into the B/CI system itself? What would the obligations of the AT to the SH be, if any? Further, what are the obligations of the SH to the AT, if any? Will the SH have a moral obligation to refrain from participating in certain activities while hosting a TS session? These aspects may be included in the SH screening process to verify 'fitness for use' certification.

Nevertheless, if a crime is committed by an SH during a TS session, which is witnessed and experienced by ATs, what should transpire? Alternatively, if a crime should happen to occur within the field of view (not involving the SH), do ATs have a moral or legal obligation to take action and report the incident? When potentially thousands of 'bystanders' or 'witnesses' are virtually present (via TS) at the scene of an ongoing crime through an SH, it may indeed change how society views the choice of non-action in some situations. The unique circumstances of TS may also raise the issue of whether these ATs might be considered witnesses in a court of law. Further, the variability of legal codes between different jurisdictions and countries would likely result in additional complexities. A set of similar moral questions are likely to arise if the SH is the victim of a personal medical emergency, natural disaster, or terrorist attack, or suffers the results of living in a conflict/war zone.

1.4.2.1 *Transcending personal and sexual frontiers; impacts on human relationships*

1.4.2.1.1 *Implications for sense of self*

The potential emergence of ubiquitously implemented B/CI systems portends intriguing possibilities for our future selves. One of the most critical foundational issues in the ontological realm (as investigated by Swan 2016) relates to how the trajectory of the classic distinct human identity might evolve once individuals are repeatedly exposed to B/CI-enabled 'cloudmind' environments. The very nature of our existence, the definition and meaning of what it is to be human, and our current modes of interaction with reality might be dramatically altered in the future. Our egocentric 'selves' appear to be forged and to emerge through the inherent capacities of consciousness and self-awareness, which have been the only known modes of lived existence for the 'classic human model.' That said, there are no steadfast justifications for the preference of the human as an organizational unit, derived either from biology or other domains. Hence, personal identity and individuality might ultimately come to be regarded as historical human artifacts.

1.4.2.1.2 Implications of virtual fully immersive sex

It is inevitable that fully immersive/fully sensorial sex will emerge as a popular B/CI application, although it may be accompanied by perhaps complex and unpredictable psychological implications insofar as human self-introspection, family/friend relationships, and the TS of alternate genders are concerned. When a TS AT inhabits an SH of the opposite sex to engage in sexual intimacies, it may be challenging to estimate the subsequent physical and psychological outcomes for the AT, who might be 'differently equipped' than the SH; hence, a requirement for supplementary sensory capacities might come into play.

Fully immersive sex may also be complicated by social challenges involving gender identity, gender expression, and sexual orientation. These concepts are often conflated and widely misunderstood, and significant discrimination and abuse continue to be directed toward members of the lesbian, gay, bisexual, transgender, and queer (LGBTQ) community worldwide. Therefore, TS sessions in which an AT's gender identity, gender expression, and/or sexual orientation differ from that of the SH imply additional complexity. Consequently, the physical and emotional impacts on ATs may also be difficult to predict in advance in this situation. The experience may challenge the preexisting beliefs of an AT regarding gender identity, gender expression, and/or sexual orientation, which may possibly lead to positive changes in tolerance for the broad variety of human sex and gender experiences.

1.4.2.1.3 Application strata

Three primary strata of B/CI applications might be sequentially developed on an escalating scale, from (1) unobtrusive resolution and enhancement of health-related pathologies to (2) fully immersive knowledge, entertainment, and experiential applications, and (3) fully fledged productive and generative cloudmind collaborations, such as interactions between multiple human minds or entities (e.g. AI/AGI/ASI) that are coordinated via the cloud/edge.

1.5 Potential sociological impacts

1.5.1 Sociological impacts of brain-embedded medical nanorobots

Increases in the adoption of technologies, our adaptation to them, and their integration into our lives and bodies appear to be progressing rapidly. According to BankMyCell, there are currently (2024) 7.1 billion smartphones in service globally (GilPres 2024). Furthermore, the World Advertising Research Center (WARC) has estimated that by 2025, nearly 3.7 billion people will access the internet using their smartphones only (Handley 2019). These figures align well with the observation that our technologies continue to shrink and become smarter, while becoming cheaper and more ubiquitous worldwide. This trend will likely be exacerbated by the recent significant advances in AI. Indeed, most mobile phone and smartphone users have come to regard these technologies as indispensable personally facilitative cognitive appendages, which will be even more solidified by the advent of powerful personalized AI agents.

Humanity's continuous exposure to increasingly advanced implantable technologies will likely eventually influence the pace of the social acceptance of brain-embedded nanomedical robots, perhaps following a similar trajectory. However, in communities with less familiarity with external and implantable devices, the level of social acceptance may be more difficult to predict. Multiple social and cultural factors are likely to impact the social acceptance of brain-embedded medical neuralnanorobots, resulting in either an increased pace of acceptance, delayed acceptance, or outright refusal of the technology. The refusal of brain-embedded nanorobots, or indeed a strong revulsion against them, might occur for several reasons. Examples include potential perceived incompatibilities with religious beliefs or the strategies employed for neural nanorobot self-ingress into the human body, whether physiologically invasive or otherwise.

1.5.1.1 Societal and interpersonal relationships

The emergence and increased implementation of mature B/CI technologies, transitioning from nascent beta trials to acceptance and possible ubiquitous use by the general public, will potentially have significant beneficial implications for individuals and human society at large. The conventional social constructs within which humans have typically interacted were severely disrupted by the global SARS-CoV-2 pandemic (declared by the World Health Organization (WHO) as a public health emergency of international concern (PHEIC) from January 30, 2020 to May 5, 2023). However, even pre-pandemic norms, which encompassed direct or travel-facilitated physical visits, business meetings, and social events/gatherings, have been steadily undergoing supplantation to a certain extent via Zoom video meetings, Skype, WhatsApp, and more. Over the next decade, due to economics or recurrent public health crises, we may witness further dramatic reductions and alterations in personal interactions in favor of real-time, fully immersive meetings in virtual environments. Indeed, the pandemic likely spurred expedited efforts toward the development of B/CI technologies. Elon Musk's Neuralink continues to make progress with its implantable brain–computer interfaces (BCIs) and obtained FDA approval in May 2023 for the investigation of its brain implants in humans (Levy *et al* 2023). These efforts stemmed from his conviction that cognition-enhancing technologies (at this nascent stage, brain–machine interfaces (BMIs)) will be necessary if humanity is ever to have a chance of keeping pace with rapidly evolving AI (Porter and Vincent 2020).

1.5.1.2 Work redefined

Considering the positions that are estimated to be replaced by AI and robotics, which may be the equivalent of 300 million full-time jobs by 2030 (Talmage-Rostron 2024), in conjunction with the gradual emergence and potentially ubiquitous adoption of B/CIs, it may be the case that a certain proportion of work tasks worldwide (contingent on the discipline/business domain) may be achieved (collaboratively or individually) via a set of standardized virtual platforms. The relationships between employees and employers will likely change, albeit perhaps not always positively. Business workers, for example, may have the capacity to create presentations and reports, or, in the scientific/academic realms, conduct experiments in virtual or remote laboratories using only their thoughts. However, taking a more abusive/dystopian/nefarious approach, employers may opt to employ a B/CI to monitor the brainwaves of their employees to glean their 'attention levels and mental states,' with the misguided aim of somehow boosting productivity to increase profits (Gonfalonieri 2020).

This would obviously seriously impinge on the rights of employees to cognitive privacy, who, it can be easily envisioned, may be let go if it is found that they mentally stray too often from the current tasks at hand. That said, by the time the first early B/CI emerges and is commercially available, all of the above tasks and a plethora of others will likely be quickly and handily taken over by AI/AGI/ASI agents that operate continuously. Thus, any type of repetitive, mundane, intensely physical labor will long be out of human hands, either being done by cloud AI agents or, in the case of physical labor, AI-imbued robots.

Where will this leave us mere mortals? Conventional anthropocentric work paradigms may be unrecognizable in ~10–20 years. With global-scale universal basic incomes (UBIs) (Gerard 2018, Gopal and Issa 2021, Evans 2023, Chang 2025), we might survive and indeed flourish in terms of personal growth, transitioning to more creative and artistic pursuits, entertainment, family life, and broader social engagement, as well as real/mixed reality travel that occupies, enriches, sustains, and enhances meaning in life.

1.5.1.3 Impacts on education and life experiences

The high-resolution, fully immersive virtual environments that will be available via a B/CI have the potential to revolutionize education and engage us in learning in ways that were never thought possible. A student living in a landlocked country with an interest in marine biology could spend time with other virtual students in an underwater research habitat on the ocean floor, gaining rare and valuable experiences. A virtual walk on the Moon or a chance to fully explore the International Space Station could inspire a young child to pursue a career in a science, technology, engineering, and mathematics (STEM) field. A group of adventurers could explore Mount Everest virtually to initially familiarize themselves with the landscape and conditions prior to making the visit in person. Budding musicians could engage with the orchestra of their choice after virtually visiting the birthplace of Mozart or virtually meeting a digital twin of the composer himself.

Students of the Japanese language could meet virtually with native speakers in Japan and become fluent far more rapidly. An artist could view the works of Monet and subsequently visit the locations that inspired these paintings. Medical education for doctors, nurses, emergency personnel, etc. could be conducted at completely new levels using fully immersive virtual training environments. The study of history could also be drastically altered if learners could visit significant places, such as Stonehenge and the Acropolis of Athens, in a fully immersive virtual environment. Indeed, one might choose to meet and engage in deep discussions with an AI-generated Albert Einstein, Martin Luther King, Abraham Lincoln, or any number of historical figures to gain insights into their personalities, concepts, thinking, and the pressing issues of their day. Learning about the lives of other living people via immersive social platforms might facilitate cross-cultural and cross-ethnic/racial understanding while promoting tolerance in a world that is desperately in need of increased empathy, basic kindness, and love. Simply put, the educational possibilities alone of a B/CI are vast, which would likely impact on every facet of society, much as the internet does today.

1.5.1.4 Travel and tourism
Totally convincing, high-resolution, fully sensorially immersive virtual environments may begin to replace physical travel and tourism, the impacts of which will translate to significantly shifting economics on a global scale. When physical travel begins to take on less significance due to the ability to visit virtual realms that are essentially indistinguishable from 'real' reality, there will likely be considerably reduced budgets allocated for face-to-face business meetings, family visits, and tourism. That said, it is reasonable to assume that many individuals will continue to crave 'real' connections and physical interactions with geographically scattered friends and loved ones. It is probable that an alluring aspect in this regard may simply relate to that refreshing 'feeling' of being offline, out of reach, and therefore free from lingering (perhaps unconscious) concerns or suspicions, for some, of being somehow surveilled or manipulated while online. Thus, the emergence of powerful B/CI technologies must be accompanied, from the outset, by verifiably unhackable, high-security, nested quantum encryption infrastructures to earn trust in this technology on the part of users, who could otherwise certainly not be blamed for non-engagement.

1.5.2 Sociological impacts of transparent shadowing

Human nature itself may produce certain side effects in some individuals following fully immersive TS sessions. Thus, it will be prudent for all SH and ATs to undergo basic psychological screening specifically as it applies to TS. It may be the case that a very small demographic of impressionable ATs might be unduly influenced or enamored by a particular SH to the point where their own personalities are temporarily altered in favor of quasi-emulating the speech or mannerisms of a particular SH they had the opportunity to briefly 'inhabit.' Typically, most individuals have one or several influential role models who personify what they perceive as desirable traits, i.e. those that they may aspire to, adopt, or emulate for themselves to a certain degree.

1.5.2.1 Privacy and security issues

The thoughts and emotions of individuals are sensitive and private and should not be accessible to any external agent, as protected by the UN Universal Declaration of Human Rights (UN General Assembly 1948). Therefore, privacy will be an extremely critical issue for prospective B/CI users; however, they may not be aware that information is being obtained from their brains. Might future B/CI technologies be used to read thoughts, gather information on the neural underpinnings of decision-making, or become the ultimate lie detector? Might psychological traits, motivations, intentions, mental states, and attitudes toward other people also be accessed? Any recordings of neural information should be held securely (by law) and guaranteed as such by the coordinating physicians and any commercial entity involved. Patients and commercial users should understand what exactly about them is being stored and for how long, as well as how it will be used. Further, they should be assured that these data will be protected from unauthorized use by international law, transgressions against which would be subject to the appropriate legal consequences. B/CI users would be required to give legal consent for any of their personal data to be shared.

The potential for hackers to access this data would threaten privacy; for example, wireless B/CI transmissions might render the system vulnerable to hacking. If certain protections of the B/CI platform are breached, hackers could potentially alter the settings (integrity) of the system, modify the activities or availability of individual or multiple neuralnanorobots, and infringe upon the confidentiality of data. To address these potential threats, nested randomly refreshed quantum encryption strategies for all transmitted data could provide robust protection and be incorporated into all B/CIs. Indeed, it is exceedingly difficult to imagine that anyone would entertain the use of these B/CI technologies if they were not verified by multiple internationally recognized top-tier medical and standards bodies as completely safe and secure from any potential nefarious breach, and such a decision could surely not be faulted.

1.5.2.2 Justice and equity of access

Initial B/CI iterations may be relatively expensive and will not have a significant market unless somehow subsidized by medical insurance or governments. This might follow the strategy employed by Tesla, Inc. where initial vehicle models were quite expensive; these, however, served to finance the development of lower-cost, mass-produced models for the public. It would indeed be inequitable and patently unjust if only the wealthy could afford B/CI technologies. Although it is conceivable that many individuals with one of the ~400 types of neurological conditions worldwide would likely benefit from B/CIs, some would argue that the considerable sums spent on developing advanced technologies such as B/CIs might be better directed to more common diseases. The contrary argument is that the alleviation of individual disabilities and suffering is a worthy endeavor. Given the already increasingly rapid pace of technological advances on a global scale, it is possible that over the next few decades (concomitant with the development of B/CI technologies), numerous classes of autonomous nanomedical devices may be developed that will have the capacity to address virtually all human diseases, including aging (Boehm 2014).

1.5.2.3 Enhancement disparities

Possibilities for the valorization and likely associated direct/indirect monetization of human augmentation will transition the B/CI beyond the realm of therapeutic devices. The enhancement of normal human cognitive or physical functionality to boost sensory, mental, and physiological endurance and capacities will undoubtedly create inequities; thus, clear preemptive regulations are warranted. Transhumanists advocate for upgrading the human body to realize capacities that may extend nominal functional levels, in some cases perhaps far beyond typical human capacity thresholds (McNamee and Edwards 2006, Hughes 2010, Porter 2017). The resultant competitive advantages might generate new forms of discrimination in society. Would this be akin to performance-enhancing drugs in sports, or might there be milder versions of enhancement that would be considered relatively innocuous and acceptable?

The expedited acquisition of new skills such as music, an additional language, or agile sports moves would likely be construed as positive outcomes all round. Having the capacity to receive information about the world instantaneously using a B/CI would essentially create a 'supermind' that is able to respond more rapidly (and perhaps in a more informed way than the normal human brain) using an internal connection to the cloud/edge. As described above, there will likely be myriad therapeutic benefits associated with first-iteration B/CI systems, where their repetitive use may stimulate/enhance neural recovery. Perhaps unwanted thoughts in depression or schizophrenia could be suppressed or eliminated using B/CIs. Conversely, it may be possible that B/CI use might adversely affect neural plasticity, which will require comprehensive investigation.

Should a B/CI be allowed to be utilized to enhance the capacities of cyborg soldiers or jet pilots by providing direct, thought-mediated links to computer/weapons systems interfaces to produce 'supersoldiers' with heightened reflexes, along with reduced decision times and action timeframes? This cyborg could be perceived as the ultimate killing machine. Many would regard the cognitively enhanced capabilities of such a soldier as immoral and unethical; however, militaries always aim for competitive advantage. Clearly, there should be a strict set of regulations established for these technologies, as is the case for various weapons systems, such as the Geneva Convention on Certain Conventional Weapons (UN 1980).

Optimistically and hopefully, humanity's future will include our maturation and evolution beyond the debauched corporate greed that exists today, as most consumer items, including healthy foods, advanced/compact power generation/battery systems, and autonomous nanomedical devices to address all that ails us, may be abundant (~10–15 years from now) due to massively distributed domestic molecular manufacturing (Factory@Home systems) (Domschke and Boehm 2014). Consequently, the military/industrial complex and the current 'profit above all' paradigm will be brought to a level where wars and corporate monopolization are roundly condemned as barbaric and obsolete relics of a pained past replete with untold suffering.

1.5.2.4 Legal considerations

A gap already exists between the current state of legislation and our capacity to augment brain function using B/CIs. Today, electroencephalogram (EEG)-based B/CI interfaces may be freely purchased, including the InteraXon device that is used to control computer games, and the Muse Headband, which is promoted as assisting 'mindfulness meditation.' Bryan Johnson's company, Kernel, developed and is commercializing 'mind reading' helmets, 'which contain nests of sensors and other electronics that measure and analyze a brain's electrical impulses and blood flow at the speed of thought, providing a window into how the organ responds to the world.' (Vance 2021). Extrapolating this, what if a company or government could monitor the brain states of millions of consumers via a B/CI? Ariel Garton (co-founder of InteraXon) founded the Center for Responsible Brainwave Technologies, which aims to prevent breaches of privacy or excessive claims for these devices. The goal is to formulate a set of standards to ensure that an individual's data is kept private and safe and that there is no misuse of this technology. These principles would extend to the B/CI as it evolves.

Technologies with the capacity for mind reading give rise to profound ethical and legal implications. A legal framework should be established that clearly differentiates between and ensures the ethical and lawful use of B/CI devices for treating disabilities and their use for augmenting human cognitive performance. The law currently recognizes the rights of the person but not of the device/system, which may lead to complexities when AI-imbued devices/systems are implanted within and are essentially considered part of the human body. And ultimately, in the case where (if) the overall B/CI system itself becomes conscious and self-aware, will we require end-user agreements? However, a quandary emerges: where does the person end and the technology begin? (Fleischfresser 2019).

There have been examples of backlashes and even physical violence against people with electronic devices attached to their bodies. Neil Harbisson attached an electronic antenna to his skull to help him with his severe color blindness. He was attacked in 2012 at a demonstration because the demonstrators thought he was filming them. Steve Mann attached a camera to his skull to allow him to access augmented-reality technology, which he overlays on the real world. He was attacked by a McDonald's employee, which is perhaps the first hate crime against a cyborg (Popper 2012). Does a user of such technology invade the privacy of other people? A visible or discovered B/CI device may encourage discrimination and stigmatization. Could individuals be denied work for fear that they might engage in industrial espionage?

1.5.2.5 Plugged-in 'slugs'? What of our physical bodies and 'real' human-to-human contact?

Humans are intensely social creatures who require consistent contact with others to remain healthy (Baumeister and Leary 1995, Carvallo and Gabriel 2006). Studies have shown that a lack of tactile contact can lead to depression and isolation and can negatively alter our biochemistry (Fisher *et al* 1976). Furthermore, a 2018 study by Cigna reported that close to half of the 20 000 adults surveyed felt that they sometimes, or always, felt alone. Close to 40% also stated that their relationships were not meaningful and that they felt isolated (Polack 2018). A meta-analysis conducted by Holt-Lunstad *et al* (2015) observed that a lack of social connections increases health risks by an amount equivalent to smoking 15 cigarettes a day: 'There is robust evidence that social isolation and loneliness significantly increase risk for premature mortality, and the magnitude of the risk exceeds that of many leading health indicators.'

One might envisage a future scenario where stadiums full of individuals are variously engaged with B/CI through virtual meetings, TS, virtual travel, entertainment, educational endeavors, and more, while comfortably reclined, albeit completely physically vulnerable. Consequently, these would have to be dedicated, highly secure B/CI facilities, likely patrolled by dedicated security personnel, humanoid robots, and/or hundreds of miniaturized drones. Realistically, any B/CI engagements should be prescheduled and time-limited, as humans can only sit still for so long (several hours) before physiological demands arise (e.g. thirst, hunger, the need to use the restroom, etc).

Furthermore, legally sanctioned B/CI integrated time constraints should be established for how long individuals may be engaged with fully immersive B/CI in their own homes, as they would again be totally physically vulnerable while they are 'gone.' In these cases, dedicated B/CI home security systems linked to certified security companies would likely be legally required. It will be prudent, with a few exceptions (e.g. emergencies, first responders, high-security situations, etc.), to incorporate automatically timed safety switches into B/CI systems to prevent individuals from 'overdosing.' This might be one or two hours at a time, with certain mandated time intervals between fully immersive B/CI sessions.

One alternate hybrid strategy to explore might be the integration of a B/CI 'toggle mode,' that switches between 'B/CI heavy' (full sensorial immersion) and 'B/CI light' (where activity is superimposed on a portion of the user's visual field with no sensorial components, akin to a heads-up display), which would allow users to carry on conversations, etc. while engaged in other activities such as eating, walking, and so on. As with today's internet usage, particularly during the multiple social constraints imposed by the global SARS-CoV-2 pandemic, there would likely be no limitations on B/CI use, the exception being its fully immersive mode.

In any case, there will undoubtedly be issues to address related to maintaining the physical well-being of short- and longer-term B/CI users. Since sedentary lifestyles are already at epidemic proportions worldwide (Arocha Rodulfo 2019, Park *et al* 2020), serious investigations into how B/CI users might maintain their physical health while becoming increasingly cognitively engaged with B/CI activities are clearly warranted.

1.6 Philosophical/ontological considerations

A comprehensive broad-spectrum philosophy of B/CIs might encompass three areas: ontology (existence and meaning), epistemology (knowledge and proof of this knowledge), and axiology (valorization, ethics, and morality).

1.6.1 Philosophical perspectives on brain-embedded medical nanorobots

In terms of humanity and personhood, several philosophical questions arise. Does the functioning B/CI become an element of the human mind, or is it just an onboard tool? Does the B/CI make the recipient more of a cyborg and less human? Does the body schema change with machine extensions? Might the B/CI induce changes in the human brain through neural plasticity, resulting in changes in personhood such that one's behavior or character is altered? In this regard, feelings of a loss of identity/control have been reported with deep brain stimulation (Yuste *et al* 2017) and implantable BCI devices (Gilbert *et al* 2019).

1.6.2 Philosophical perspectives on transparent shadowing

The ability to literally 'inhabit' and walk in someone else's shoes in real time with full sensorial resolution may become a reality with the advent of B/CI technology and engagement with its envisaged TS application. From a philosophical standpoint, this may be the epitome of facultative means through which humans may more fully empathize with their fellow beings. If we have the capacity to actually 'feel' what it is like to be starving on the streets, harshly discriminated against, or riddled with debilitating diseases, we may learn to be far humbler and more compassionate toward others. We might more poignantly and clearly realize that all of us (without exception, as none of us is perfect) deal with personal struggles, doubts, fears, insecurities, etc. For the first time in human history, we may collectively have the opportunity to rise above the divisiveness and outright hate that we seemingly witness more every day on a global scale. Cumulatively, this would amount to a significant positive step in our evolution as a species and may indeed be an essential factor for our long-term survival.

1.6.3 Impacts on human civilization—TS as a gateway to a ten-billion-synapse world mind?

In discussions that involve the coalescence of individual human minds, the concepts of a hive mind and the Borg Collective (from the *Star Trek* series) often arise. The concern is that human civilization may be heading in this direction if this type of brain-linking technology is ubiquitously implemented. While the future cannot, of course, be predicted with certainty, a negative version of a collective consciousness is not the only possible model for humanity's future. Science fiction has not provided us with only negative examples to consider. A much more positive conceptual design has been presented in James Cameron's film, *Avatar*, in which the infrastructure of the entire planet (Pandora) integrates a world mind that the inhabitants can plug into and communicate through. It is obvious that humanity should strive to direct the evolution of this envisioned novel technology toward positive results, as it may likely serve as a gateway to the further evolution of human civilization. As mentioned above, this may ultimately involve our seamless integration with our technologies.

1.7 Higher planes of existence

There are likely to be unanticipated psychological and emotional impacts that result from joining individual human minds en masse via B/CIs. Such impacts are indeed difficult, if not impossible, to predict in advance. However, one may ponder whether, at some juncture, B/CI users will begin to feel more connected to and comfortable with humanity as a whole because of this process. The feeling of oneness with humanity may be akin to what many astronauts experience while viewing our planet from space, referred to as the 'overview effect.' These astronauts report life-changing experiences that include an increased awareness of their connection to all of humanity and the Earth. It has been suggested that if everyone on the planet could experience the overview effect from space, several trends would occur, such as increased global citizenship and social investment in people (Okushi and Dudley-Flores 2007). If a similar overview effect were to occur in B/CI users as a result of being connected to so many other human minds, the potential would exist to initiate similar changes on a global scale. If manifested, the positive impacts on human civilization would be highly significant.

1.7.1 Brain/cloud-interface-facilitated access to universal energy fields and spiritual realms

There will undoubtedly be many who will seek to employ a B/CI to further enhance their religious, spiritual, or enlightenment experiences and connections with these realms. Yet others will yearn to more intensely plumb the depths of the posited unified field (as in transcendental meditation) and universal intelligence, etc. Furthermore, more esoteric and exotic endeavors might encompass extrasensory perception (ESP), remote viewing, mind melding, directed thought energy, etc. for personal growth, the benefit of others, the planet, and beyond. In any case, the full range of cognitive adventures will likely know no bounds.

1.8 Requesting cognitive quiet

It is certainly conceivable, and will undoubtedly be inevitable, that at some juncture for any given B/CI user, it might all just become too much, as fully immersive cognitive/intellectual overstimulation may leave some feeling quite overwhelmed, mentally and emotionally drained, and/or burned out. Thus, there would most definitely be recommended limits for the duration of B/CI engagement sessions, as well as robust integrated disengagement protocols, whereby any B/CI-enabled individual could immediately extricate him/herself from connectivity at any time. Another, more stringent and equally robust set of protocols would be established for those who may elect, for any reason, to auto-purge their brain of the resident B/CI prosthesis altogether.

Furthermore, a B/CI user might possibly develop more severe mental health-related problems, as has been suggested to be the case for other brain interfacing technologies (Trimper *et al* 2014). Immersive experiences that are frightening or upsetting may trigger feelings of anxiety or depression in some users. Also, exposure to situations involving violence or other traumatic experiences may even result in a form of post-traumatic stress disorder (PTSD) in B/CI users. In addition, certain experiences could possibly trigger symptoms if the B/CI user has a preexisting mental health condition, including depression, anxiety, or PTSD. Comprehensive screening and stringent protocols would need to be established during the development of the B/CI platform in consultation with mental health professionals to properly address these concerns. As a medical waiver will likely be required at the time of accepting and engaging a B/CI, options for obtaining immediate mental healthcare (which may be channeled through the B/CI itself via an avatar, AI agent, or a direct live link to a human psychologist) could be presented to users at that time. Interestingly, while a B/CI may, in some cases, potentially create or exacerbate mental health issues, there will also be the potential for reducing or perhaps even eliminating distressing symptoms.

The immersive experience itself may result in increased positive feelings (such as happiness and relaxation) and therefore play a therapeutic role for users. The user may opt for certain experiences or even specially designed programs (such as guided meditation) when feeling overwhelmed or burned out in physical life. At the clinical level, other types of BCI technologies are already being used to modify unhelpful thought and behavior patterns in patients with severe mental health problems (Ruiz *et al* 2013); therefore, this may be a promising area for future development. B/CI technologies may have the potential to manage a variety of mental health challenges that have historically been difficult to treat, which could result in significantly positive impacts on the quality of life for countless individuals worldwide.

1.9 Conclusions

The emergence of a hypothetical (for now) nanomedically enabled B/CI within the next few decades clearly raises ethical and moral issues relating to autonomy, agency, identity, consent, and human enhancement. The conscious transfer of information between human brains at a distance is referred to as 'hyperinteraction' (Grau *et al* 2014). Today, existing ethics guidelines do not, and most likely cannot, comprehensively address all the perceived issues and challenges presented as relating to a B/CI, not to mention the myriad unforeseen developments or scenarios that might arise concurrently with these potentially immensely powerful technologies. Nevertheless, we propose that the time is now ripe (decades ahead of the potential emergence of such technologies) for world bodies such as the United Nations (UN) (United Nations Educational, Scientific and Cultural Organization (UNESCO), the Office of the High Commissioner for Human Rights (OHCHR), the UN Human Rights Council); the WHO (Ethics and Health Unit); the World Medical Association (WMA) (Declaration of Helsinki on Medical Research Ethics); the International Bioethics Committee (IBC) (UNESCO advisory body); the Council of Europe (European Convention on Human Rights, Oviedo Convention on Bioethics); the IEEE Global Initiative on the Ethics of Autonomous and Intelligent Systems; Partnership on AI; the European Commission's High-Level Expert Group on AI; the International Association for Ethics Education (IAEE); and the Global Ethics Network (Microsoft Copilot 2025) to begin to consider, discuss, and address these challenges and for governments to develop preliminary preemptive regulatory policy frameworks and legislation (Yuste *et al* 2017). This would allow the positive aspects of B/CI technologies to be optimized and maximized for the benefit of humanity while minimizing and ideally negating the converse.

Acknowledgments

We acknowledge here that with recent rapid advances in artificial intelligence (AI), quantum computing (QC), and the likely downstream emergence of AGI and ASI, that a B/CI might be achieved completely non-invasively using technologies (perhaps employing energy, frequency, and vibration, as envisioned by Nikola Tesla) that we cannot even imagine at present.

Parts of this chapter have been reproduced with permission from Rosenfeld and Broekman (2020).

Appendices

Brain/cloud interface queries

1) *How do you perceive that people might feel about the presence of perhaps many billions of autonomous nanorobotic devices within their neocortexes (interfaced with targeted neurons and synapses), which would allow them to have (at will) instantaneous access to any facet of human knowledge ever digitized, as well as to experience fully immersive movies, games, and the possibility for 'transparent shadowing'? (note: the self-installation of these nanorobotic entities would be completely reversible at any juncture)*

Robin Farmanfarmaian (RF)

https://twitter.com/Robinff3

We'll see early adopters, but I think there will be huge fear around this as well that will potentially cause a large amount of political intervention and regulation. Imagining a world where we're past all that and this is a normal thing to do, it will alter reality so significantly, as to be completely indistinguishable from what our reality is today. When are you inside VR or out of it? What is your real body, and which one is the VR body? The physical world will become indistinguishable from the virtual world. Already we've seen you can make money in the virtual world, as people have been doing for years inside of Second Life.

Denice Lewis (DL)

https://www.linkedin.com/in/denicedlewis/

That depends on how you define artistic endeavor. In my opinion there exists artistry in all forms of creativity including practical creativity such as mathematics, justice, relationships of all potential as long as the creative process is possible through communication without the requirement of physical presence. For example....painting. How would one collaborate with another entity and produce in real time a physical result as is a painting??? To me it is possible to compose.... but that is simply the ability to share a formula created by the participation of multiple entities with an outcome of a single result that could be read and then translated based on the information shared. How would that be possible to succeed when physical items are necessary to accomplish a finished piece such as a painting? Perhaps a digital version could work but that wouldn't result in an organically realized result. Much less in real time.

2) *How do you suppose that people's perception of reality, and 'real' human interaction might be changed, if at all, following their exposure to B/CI for a certain period of time (say a few months)?*

RF: People's perception of reality will alter dramatically, even after just a short period of time, if they live in part inside a virtual world with a BCI. The 2 worlds will become indistinguishable.

DL: That would depend on a myriad of components which would be dictated per each individual who participates. Personality, education, geographic influences and socio-economic considerations etc would factor into that result. I do, however, think that the world is now at a point where there is far greater probability of receptivity given the current state of technological advances that have occurred within the past 20 years

3) *How might B/CI technologies impact society, psychologically, as a whole? For instance, in view of how the Internet has already altered the way, and frequency with which we interact with each other, will people find that they may further distance themselves from, or don't find it necessary to engage in, 'real' interactions with others, as it may be increasingly difficult (due to the ultrafine resolution of fully experientially immersive experience provided by the B/CI) to distinguish between virtual and real worlds? (note: I suppose that there could be an omnipresent icon somewhere in the users field of view, whenever they are in B/CI mode, as a reminder and verification that they are indeed immersed within a virtual world)*

RF: People will potentially start to live inside the VR world more than the physical world—and why not? You can alter yourself to look any way you'd like. You can change your voice, way you walk, sex, anything. You don't need to expend effort driving in traffic or waste time traveling between places, you are limited to your physical form—you are only limited by your, and others, imagination. Why fly to Africa, get live vaccinations, experience the time shift, risk of clean water, food change, and spend a large amount of money to visit, when instead, through your B/CI, you are there instantaneously, without risk, without the uncomfortable, long plane ride, packing, shots, and all the rest of the hassle? In fact, most people aren't going to be able to physically visit a lot of places in the world, but with the virtual world, they can visit every country in the world.

DL: I think that you have answered your own question here. How does one define 'real'. I personally have 'real' friend due to social media whose physical body has not yet been experienced by me. The beauty of the Internet is that it affords each of us to connect in a very real state of being as our intellect, personality, patience etc are experienced truthfully and in 'real' time. Our efforts here in this instance is indicative of that.

> 4) *In the spiritual realm, might B/CI have the capacity to facilitate more intense spiritual awareness, communal worship, and the attainment of bliss states, as well as to enhance the feeling of the oneness of humanity?*

DL: Indeed! Energy isn't always measurable by the human eye but it does exist as real as anything else. There is scientific data that proves beyond a reasonable doubt that water is a living entity and can be observed and measured based upon its own experience. It is possible to pray over water and witness its evolution after by freezing it and the looking microscopically at its crystallization process that reveals the result of the experience the water has incurred. Water that has been exposed to negativity has a very different reaction that is seen physically as well. We are all connected to each other via many perspectives. Spirituality is one way to recognize this. I wrote a little quote years ago that I will share with you affirming what I believe to be true.—'We are all facets of the diamond of life reflecting the light of the One most high.' ~ Denice D. Lewis

> 5) *If you were responsible for formulating policies for the safe and secure use of B/CI technologies, what would be your three most important recommendations?*

RF: This would be dependent on how advanced/what features the BCI has. Assuming we can control other people's bodies (as has already been done in a lab between 2 scientists, one scientist moved the other's finder by thinking.)

1) Just as laws exist that (other than law enforcement, imminent risk of injury, or medical intervention), under normal circumstances, you can't control another human being's movements without permission. This will get tricky to detail, as when is it medically necessary? When someone is about to jump off a bridge, or accidentally walk into traffic?
2) Privacy. There must be some barriers that keep private information and thoughts secure.
3) All laws that required physical intervention be upheld and applied to BCIs. Meaning, ou are liable when you use a BCI to control something else that breaks a law.

DL:

1) Be responsible for your words, thoughts, and intentions. They manifest into realty that can and does result in action and reaction.
2) Be available 24/7 to assist in situations where participants may not be as spiritually evolved as in the Ayahuasca ceremonies for example. Should a problem arise an 'elder' should always be present or immediately available.
3) Be compassionate and considerate of the experiencers emotions. Never deny their right to their emotional state after such an experience. Always endeavor to direct the experience to a positive outcome no matter what the experience is.

Keirsten Snover

https://www.linkedin.com/in/keirstensnover/

1) The establishment of a multidisciplinary B/CI-ELSI Working Group. (Ethical, Legal, & Social Implications) Possible tasks could include: connecting with interested professionals; collecting related information, resources, and research; formulating policy and position statements; creating relationships with stakeholder organizations; attending and/or presenting at relevant conferences and workshops; conducting original research.

2) Formal qualitative assessments in multiple cultural settings, in order to better understand the impact of B/CI that will further connect a culturally diverse world. These assessments may include: risks and benefits in various populations, factors affecting use or rejection, cultural perspectives surrounding the introduction of the technology, and other topics.

3) Special consideration should be given to how historically marginalized groups may encounter and experience the B/CI technology. This would include how B/CI may impact these groups differently and how any potential barriers to their participation could be removed. (Members o of marginalized groups may vary by region, but often include those with disabilities, those of minority ethnic/racial groups, the LGBTQ community, women, and those with low socio-economic status.)

B/CI perspectives

A number of intriguing perspectives have been offered by several individuals from diverse backgrounds in response to the set of queries below. This comprised an attempt to glean some insights into the potential attitudes that may accompany the potential emergence of B/CI's from a cross-section of society. Viewpoints such as these will be valuable consideration that may likely facilitate the formulation of prudent policies to guide the potential development and implementation of these unprecedented technologies in the future.

Cynthia Duval

https://www.linkedin.com/in/cynthiaduval/

We can expect nanomedical technology like almost any other innovation, to have both positive and negative social, physical *and* psychological effects. There is something deeply creepy about inserting billions of nanodevices into our bodies and perhaps most especially our brains and skulls. Would I do this to myself for the benefit of being viscerally connected to information on the Internet? This is unimaginable. But to save my life or my son's life? That opens up a world of possible acceptance. What these devices look like and the ingress/egress conversation will frighten people. Showing their actual size, making them seem softer, less prone to teaching delicate flesh, will help some. Transparent Shadowing? It might make for fun experimentation and appeal to fringe scientists but people will likely question the benefit for years to come. It would be a very interesting concept to explore in a movie). I am personally quite pleased to see your two abstracts presented together as all too often the forward motion of invention is so powerful it steam rolls over in-depth consideration and anticipation of the unintended consequences of invention. We move forward so quickly that we fail to prevent the unintended consequences, opting instead to fix them if and when they happen. We all know how poorly and on what a large scale this scenario plays out.

Robin Farmanfarmaian

https://twitter.com/Robinff3

There are many people already with implanted BCIs, especially when you consider the number with cochlear implants alone (over 800 000 estimated for 2025). Imagine a world where instead of taking a medication, you have a chip implanted that controls behavior, addiction, seizures, neurological disorders? Beyond using them to control objects, robots, games, travel, maybe we start using them for sex. A BCI could fully simulate a sexual encounter, complete with full body responses, all without you taking the risk of striking out with a potential partner using a bad pickup line.

Joanna Jaoudie

https://www.linkedin.com/in/joannajaoudie/

https://www.linkedin.com/in/joannajaoudie/ It is logical that one would question how a successful B/CI is to be physically construed for adequate use. Two conceptual nanomedical strategies are proposed: a physiological versus a wireless model. The physiological model asserts the use of autonomous nanorobots in the millions swimming around our most delicate neocortex, supposedly invading the very layers that control our thoughts, desires and intentions. Each nanorobot comes equipped with the capacity to attach, and therefore communicate, however briefly, with various neurons in our brain, essentially facilitating the characteristic output that occurs upon that interaction-much like a drug. I would imagine that these nanobots (as I like to call them) would be directly injected/implanted into our brains, running the risk of our immune system rejecting its new foreign host. The supposed less invasive equivalent to this is a cranially embedded wireless model; although, for all intents and purposes-anything that requires even the slightest surgical procedure would feel invasive enough. Here, I am addressing physical invasiveness.

Anything that runs the risk of being externally hacked into will be viewed as invasive on a mental level, and that can be viewed as invasive to the highest degree. Once in our brains, in our thoughts. Once in our thoughts, the body becomes but a vessel driven by thought. People throughout the ages have already worried about the impact that subliminal messaging and advertising has brought. There are existing brainwashing techniques as well. We are constantly being monitored. At least these methods exist externally to our physical bodies. Here we are talking about permitting the direct entry of these forces into our physical bodies, allowing them to penetrate our mental structures, and presumably have the power to alter how much knowledge we are empowered by. Who is to guarantee the direction in which this power should be used? Will the technologically savvy, adept academic, or the person with higher intellectual scores perform better and/or more responsibly in reaction to using this technology? More than the average consumer? Will each person have equal rights to utilize this technology or will it be inversely related to a person's socio-economic status? How much right is one granting the use of their privately held insights during Transparent Shadowing?

When talking about the aforementioned statement that 'once in our brains, now in our thoughts', I'm also opening doors to ponder about higher 'spiritual' realms as well, to investigate the long philosophical debate of whether or not our thoughts are attached to our 'souls'—not just our minds—and even the possibility that our minds and souls are one, not necessarily physically attached to matter like the brain, but finding an ideal form of physical expressionism and manifestation within our human race. To me, that is an optimistic answer to the question of how it is one can even tap into these higher realms. We are unable to provide evidence that such realms are physically housed in the vast existence of our universe as scientists, and the wild notion that we should go hunting for intergalactic portals (for instance) definitely seems like a waste of our time (while also assuming that a portal lacks the physical parameters needed to locate one). Like many things in life, there is a starting point for events that can usually be temporally pin pointed with, say, longitudinal/ latitudinal references-data points of sorts. If we can't physically pin point this data externally, we have to start looking within—why not quite literally within the brain? Why shouldn't the brain be the physical gateway that houses a universal concept like consciousness? It is simultaneously the object that makes us believe we are all made of the same stuff, while at the same time insistent that we are quite different from each other.

This consciousness cannot be defined and restricted within parameters. That is why it cannot be found in one place but is an amalgamation of experiences passing through multiple brains and life matter. Consciousness better mirrors the Bergsonian argument that 'life possesses an inherent creative impulse which creates new forms as life seeks to impose itself on matter' (Bergson). It is the stroke of genius that is actualized and captured for a mere moment and then lost and found again in the form of another genius revelation. It comes and goes in waves and has purpose.

We have a desire to know each other and to learn what makes us better get ahead in our own game at life. We are brought up to believe that education is essential to the process of excelling. We may have mentors or people who we regard as highly intellectual, who offer insights that provide added value to our own lives. It could very well be that one aspires to be like their mentor. Now what if they could actually simulate an episode of that person's life and character, to directly experience stepping into that person's shoes, or become that person's shadow for a while? Will this person's skills successfully be imparted upon the attendee?

Heard vocal instructions of the host are used to guide and satisfy the attendee's temporal and sensual experience of the host, but all other access to host's thoughts, emotions and self-speak are blocked to attendee. Yet, how can insights truly be imparted and a full sensual experience of the host be had by the attendee if the attendee can only access what is being externally communicated to the host, not what host feels (a result of emotion), thinks (a result of thought) or internally vocalizes (self-speak)? Vocalizing our thoughts (self-speak) is vital to the thought process, particularly in decision-making. It is also central to introspective speak, and paramount in formulating a self-understanding of how we put together information/ behavioral stimuli gathered from a range of external (environmental) and internal (individualized) configurations. One tends to be more liberal and uninhibited with self-speak, so it's not a stretch to assume that therein lies our darkest and most formidable truths as well. This is vital information you would be eliminating from Transparent Shadowing.

Nevertheless, preserving the host's privacy and rights is essential. We would have to withhold sensitive data that would unlock the true potential of the host, or else, who would own your intellectual property? How do you protect your information? Are we going to have to start copyrighting our thoughts now? Where do we draw the line? Privacy and ownership have never been so fragile.

We seek to be connected. It is the joy of being part of the same network— a communal feeling or oneness in an activity, belief or faith that makes feel a sense of belonging. We are delighted when we encounter people who are like us, or who we can share common views with because it reinforces and strengthens our thoughts with positivism. We often gather in physical spaces to celebrate traditions and concepts. Who is to say that the same cannot be done in a virtual connected cloudmind that can break the barriers of traditional physical space and really foster a greater sensation of united oneness? It is likewise delightful to meet diverse people who have something different and new to offer and teach us. I would not be surprised if by being exposed to new perspectives, comfort and relatability is found, and chaos is quietened.

Current technologies offer several ways for us to communicate with each other, intentionally from a distance. We network in a multitude of ways, but we have never been more physically disconnected. The implications that more immersive techniques and technologies can have on us go well beyond today's worries of how much screen time we allow each other to have in the same room with each other while we share a meal together. They extend to how long it'll take before we start to question if we even need to be in the same room together ever. We cannot take for granted who we are and how important it is to remain physically connected.

References and further reading

Appelbaum P S 2007 Clinical practice. Assessment of patients' competence to consent to treatment. *N. Engl. J. Med.* **357** 1834–40

Arocha Rodulfo J I 2019 Sedentary lifestyle a disease from xxi century. *Clin. Investig. Arterioscler.* **31** 233–40

Baumeister R F and Leary M R 1995 The need to belong: desire for interpersonal attachments as a fundamental human motivation. *Psychol. Bull.* **117** 497–529

Bear G 1997 *Slant (Queen of Angels #4)* (New York, NY: Open Road Media)

Biroudian S, Abbasi M and Kiani M 2019 Theoretical and practical principles on nanoethics: a narrative review article. *Iran. J. Public Health.* **48** 1760–7

BitInfoCharts 2025 https://bitinfocharts.com/bitcoin/ (accessed 15 July 2025)

Boehm F J 2014 *Nanomedical Device and Systems Design—Challenges, Possibilities, Visions* (Boca Raton, FL: CRC Press)

Brin D 1998 *The transparent society: will technology force us to choose between privacy and freedom?* (Reading, MA: Addison-Wesley)

Buchanan A 2011 *Better than Human: The Promise and Perils of Enhancing Ourselves* (Oxford: Oxford University Press)

Buchanan A 2013 *Beyond Humanity: The Ethics of Biomedical Enhancement* (Oxford: Oxford University Press)

Carvallo M and Gabriel S 2006 No man is an island: the need to belong and dismissing avoidant attachment style. *Pers. Soc. Psychol. Bull.* **32** 697–709

Chang W C 2025 A grand plan for health equity: philosophy of health equity. *Int. J. Equity Health.* **24** 183

Clausen J, Fetz E, Donoghue J, Ushiba J, Spörhase U, Chandler J, Birbaumer N and Soekadar S R 2017 Help, hope, and hype: ethical dimensions of neuroprosthetics. *Science* **356** 1338–9

Deleuze G and Guattari F 1972 *Anti-Oedipus: Capitalism and Schizophrenia* (Viking Penguin)

Domschke A and Boehm F J 2014 (Essay in response to the question, how should humanity steer the future?)—quandary—are molecularly manufactured burgers imbued with the life force? FQXi Foundational Questions Institute https://forums.fqxi.org/d/2085-quandary-are-molecularly-manufactured-burgers-imbued-with-the-life-force-by-frank-josef-boehm-and-angelika-domschke (accessed 17 July 2025)

EMC Corporation 2012 New digital universe study reveals big data gap: less than 1% of world's data is analysed; less than 20% is protected: study *PR Newswire* https://prnewswire.com/news-releases/new-digital-universe-study-reveals-big-data-gap-less-than-1-of-worlds-data-is-analyzed-less-than-20-is-protected-183025311.html (accessed 24 April 2024)

Evans E J 2023 Universal basic income: a synopsis for social work. *Health Soc. Work.* **48** 7–10

Fisher J D, Rytting M and Heslin R 1976 Hands touching hands: affective and evaluative effects of an interpersonal touch. *Sociometry* **39** 416–21

Fleischfresser S 2019 Legal problems loom for cyborgs *Cosmos* https://cosmosmagazine.com/technology/robotics/humans-machines-and-lawyers-legal-problems-loom-for-cyborgs/ (accessed 28 April 2024)

Foucault M 1977 *Discipline and Punish: the Birth of the Prison* (Pantheon Books)

Friedrich E V, Suttie N, Sivanathan A, Lim T, Louchart S and Pineda J A 2014 Brain-computer interface game applications for combined neurofeedback and biofeedback treatment for children on the autism spectrum. *Front. Neuroeng.* **7** 21

Gerard N 2018 Universal healthcare and universal basic income. *J. Health Organ. Manag.* **32** 394–401

Gilbert F, Cook M, O'Brien T and Illes J 2019 Embodiment and estrangement: results from a first-in-human 'Intelligent BCI' trial *Sci. Eng. Ethics.* **25** 83–96

GilPres 2024 How many people own smartphones? (2024–2029). what'sthebigdata 2024 https://whatsthebigdata.com/smartphone-stats/ (accessed 27 April 2024)

Gonfalonieri A 2020 What brain-computer interfaces could mean for the future of work. *Harv. Bus. Rev.* https://hbr.org/2020/10/what-brain-computer-interfaces-could-mean-for-the-future-of-work (accessed 13 April 2024)

Gopal D P and Issa R 2021 What is needed for Universal Basic Income? *Br. J. Gen. Pract.* **71** 398

Grau C, Ginhoux R, Riera A, Nguyen T L, Chauvat H, Berg M, Amengual J L, Pascual-Leone A and Ruffini G 2014 Conscious brain-to-brain communication in humans using non-invasive technologies. *PLoS One* **9** e105225

Grunwald A 2010 From speculative nanoethics to explorative philosophy of nanotechnology. *Nanoethics* **4** 91–101

Handley L 2019 Nearly three quarters of the world will use just their smartphones to access the internet by 2025 *CNBC* https://cnbc.com/2019/01/24/smartphones-72percent-of-people-will-use-only-mobile-for-internet-by-2025.html (accessed 28 August 2020)

Harman M 2023 *Please Stop Trying to 'Fix' My Disability* (Mighty Proud Media, Inc.) https://themighty.com/topic/disability/people-with-disabilities-dont-need-to-be-fixed-or-cured/ (accessed 27 August 2023)

Holt-Lunstad J, Smith T B, Baker M, Harris T and Stephenson D 2015 Loneliness and social isolation as risk factors for mortality: a meta-analytic review. *Perspect. Psychol. Sci.* **10** 227–37

Hughes J 2010 Contradictions from the enlightenment roots of transhumanism. *J. Med. Philos.* **35** 622–40

Jaarsma P and Welin S 2012 Autism as a natural human variation: reflections on the claims of the neurodiversity movement. *Health Care Anal* **20** 20–30

Klein E 2016 Informed consent in implantable BCI research: identifying risks and exploring meaning. *Sci. Eng. Ethics.* **22** 1299–317

Lane F J, Huyck M, Troyk P and Schug K 2012 Responses of potential users to the intracortical visual prosthesis: final themes from the analysis of focus group data. *Disabil. Rehabil. Assist. Technol.* **7** 304–13

Martinazzi S and Flori A 2020 The evolving topology of the Lightning Network: centralization, efficiency, robustness, synchronization, and anonymity. *PLoS One* **15** e0225966

Martins N R B *et al* 2019 Human brain/cloud interface. *Front. Neurosci.* **13** 112

McNamee M J and Edwards S D 2006 Transhumanism, medical technology and slippery slopes. *J. Med. Ethics.* **32** 513–8

Merleau-Ponty M 1945 *Phénoménologie de la Perception.* (Librairie Gallimard)

Microsoft Copilot 2025 List of world bodies dealing with human ethics and morality. [derived from AI-generated content]. Personal communication

Musk E 2019 Neuralink. An integrated brain–machine interface platform with thousands of channels. *J. Med. Internet. Res.* **21** e16194

Okushi J and Dudley-Flores M 2007 Space and perceptions of space in spacecraft: an astrosociological perspective. *American Institute of Aeronautics and Astronautics. AIAA 2007-6069. AIAA SPACE 2007 Conf. & Exposition (Long Beach, California, 18–20 September 2007)* (Reston, VA: American Institute of Aeronautics and Astronautics, Inc.)

Park J H, Moon J H, Kim H J, Kong M H and Oh Y H 2020 Sedentary lifestyle: overview of updated evidence of potential health risks. *Korean J. Fam. Med.* **41** 365–73

Polack E 2018 New cigna study reveals loneliness at epidemic levels in America. https://prnewswire.com/news-releases/new-cigna-study-reveals-loneliness-at-epidemic-levels-in-america-300639747.html (accessed 29 April 2024)

Popper B 2012 New evidence emerges in alleged assault on cyborg at Paris McDonald's https://theverge.com/2012/7/19/3169889/steve-mann-cyborgassault-mcdonalds-eyetap-paris (accessed 29 April 2024)

Porter A 2017 Bioethics and transhumanism. *J. Med. Philos.* **42** 237–60

Porter J and Vincent J 2020 Elon Musk promises demo of a working Neuralink device on Friday *The Verge* https://theverge.com/2020/8/26/21402240/neuralink-august-2020-event-brain-machine-interface-working-demonstration (accessed 26 August 2020)

Rosenfeld J V and Broekman M 2020 Brain-machine interface technology in neurosurgery *Ethics in Neurosurgical Practice* ed S Hoeybul (Cambridge: Cambridge University Press) **22** 224–33

Roskies A L 2015 Agency and intervention. *Philos. Trans. R. Soc. Lond. B. Biol. Sci.* **370** 20140215

Ruiz S, Birbaumer N and Sitaram R 2013 Abnormal neural connectivity in schizophrenia and fMRI-brain-computer interface as a potential therapeutic approach. *Front. Psychiatry* **4** 17

Scroxton A 2018 Medical firm debuts internet-connected prosthetic limbs *Computer Weekly* https://computerweekly.com/news/252448904/Medical-firm-debuts-internet-connected-prosthetic-limbs (accessed 24 April 2024)

Levy R, Taylor M and Sharma A 2023 Elon Musk's Neuralink says has FDA approval for study of brain implants in humans. *Reuters* https://reuters.com/science/elon-musks-neuralink-gets-us-fda-approval-human-clinical-study-brain-implants-2023-05-25/ (accessed 29 August 2023)

Shih J J, Krusienski D J and Wolpaw J R 2012 Brain-computer interfaces in medicine. *Mayo Clin. Proc.* **87** 268–79

Sunshine J C and Paller A S 2019 Which nanobasics should be taught in medical schools? *AMA J. Ethics.* **21** E337–346

Swan M 2015a Machine ethics interfaces: an ethics of perception of nanocognition *Rethinking Machine Ethics in the Age of Ubiquitous Technology* ed J White (London: IGI Global) 97–123

Swan M 2015b *Blockchain: Blueprint for a New Economy* (Sebastopol CA: O'Reilly Media)

Swan M 2016 The future of brain-computer interfaces: blockchaining your way into a cloudmind *J. Evol. Technol.* **26** 60–81

Talmage-Rostron M 2024 How will artificial intelligence affect jobs 2024–2030. Nexford university. January 09, 2024. https://nexford.edu/insights/how-will-ai-affect-jobs (accessed 27 April 2024)

Trimper J B, Wolpe P R and Rommelfanger K S 2014 When 'I' becomes 'We': ethical implications of emerging brain-to-brain interfacing technologies. *Front. Neuroeng.* **7** 4

United Nations General Assembly 1948 *Universal Declaration of Human Rights (217 [III] A)* (Paris) https://un.org/en/about-us/universal-declaration-of-human-rights (accessed 29 August 2023)

United Nations 1980 2. Convention on Prohibitions or Restrictions on the Use of Certain Conventional Weapons which may be deemed to be Excessively Injurious or to have Indiscriminate Effects (with Protocols I, II and III) Treaty Series 1342 *(New York, NY: United Nations Office of Legal Affairs)* XXVI–137 *https://treaties.un.org/pages/ViewDetails. aspx?chapter=26&clang=_en&mtdsg_no=XXVI-2&src=TREATY (accessed 25 September 2025)*

Vance A 2021 Kernel helmet that is claimed to read human mind starts shipping for $50,000 in US *Bloomberg* https://gadgets360.com/wearables/news/kernel-helmet-price-usd-50000-read-human-mind-analyse-brain-ceo-bryan-johnson-2466256 (accessed 30 August 2023)

Williams S 2017 Optogenetic therapies move closer to clinical use *The Scientist* https://the-scientist.com/optogenetic-therapies-move-closer-to-clinical-use-30611 (accessed 24 April 2024)

Yuste R, Goering S, Arcas B A Y, Bi G, Carmena J M, Carter A *et al* 2017 Four ethical priorities for neurotechnologies and AI. *Nature* **551** 159–63

Zao J K, Gan T T, You C K, Chung C E, Wang Y T, Rodríguez Méndez S J *et al* 2014 Pervasive brain monitoring and data sharing based on multi-tier distributed computing and linked data technology *Front. Hum. Neurosci.* **8** 370

Chapter 2

Analysis of power, locomotion, communications, and navigation for microbots in the brain

Tad Hogg

The realization of the full potential of a brain/machine interface (BMI), brain/computer interface (BCI), and a hypothetical nanomedically enabled brain/cloud (edge) interface (B/CI) will require microscopic devices that connect to and individually interact with myriad neurons and synapses. Such interfaces will necessitate the deployment of multitudes of microscale devices that are small enough to access and interact with neurons on the millisecond timescales of their signal transmissions. These requirements mean that the devices must autonomously sense and perform tasks; i.e. the devices will be microscopic microbots (~ 1 μm in diameter). The establishment of effective and safe interfaces with neurons will require several robotic capabilities, including extracting power, safely traversing tissues, communicating with other robots to coordinate activities, and orienting themselves with respect to their environments. This chapter describes several techniques that might enable these capacities. Due to their diminutive size and large numbers, these robots must be capable of determining and implementing their actions with, at most, a coarse level of external guidance. Thus, the focus of this chapter is on short-range activities that take place over distances comparable to the dimensions of the robots themselves and the cells they may interact with. In addition to the evaluation of individual capacities, several design trade-offs will be articulated that arise from the limited spatial volume, surface area, and computational capabilities of these microscopic robots.

2.1 Introduction

Current BMI and BCI technologies employ external electrodes applied to the scalp or implanted electrodes (Shih *et al* 2012). These interfaces provide multiple channels

doi:10.1088/978-0-7503-2144-0ch2
2-1

that monitor and respond to neural signals (O'Doherty *et al* 2011, Collinger *et al* 2013). External electrodes are relatively simple to use but cannot monitor and send signals to individual neurons. Alternatively, implants connect directly to neurons but are currently limited to relatively small numbers of electrodes and are invasive, thus increasing the likelihood of infections if they are tethered to external power sources via wires that emanate from the body.

Advances in the miniaturization of implantable medical devices will enable increasingly capable BMIs and BCIs (Llinás *et al* 2005). This improvement is one example of the medical benefits of advanced nanotechnology (Morris 2001, Betancourt and Brannon-Peppas 2006, Thomas *et al* 2007, Monroe 2009, Sánchez and Pumera 2009). This technology is evolving the ability to manipulate tiny objects (e.g. submillimeter polymer-based robots that can grasp and manipulate micronscale objects over a range of ~100 μm) (Jager *et al* 2000), and at molecular scales, DNA-based devices can precisely move cargo based on binding-site recognition (Thubagere *et al* 2017).

Further abilities arise from combining the precision of nanoscale devices with the programmability currently available only in larger machines. This combination is the basis for microscopic robots (microbots) (Freitas 1999, Nelson *et al* 2010, Boehm 2013), each of which is small enough to access and interact with single cells in the body. Advanced applications for microbots include currently hypothetical nanomedically enabled BMIs and a prospective, even more sophisticated B/CI, which would be enabled by several classes of 'neuralnanorobots' (Martins *et al* 2019). While these complex technologies remain well beyond the realm of current knowledge and practical fabrication, a quantitative evaluation of the plausible applications of microbots can assist in identifying options on the pathway toward their ultimate realization.

For example, the data-processing capacities of these robots will be particularly useful for BCIs, as they will enable the computation of rapid responses to signals from individual cells without the latency involved in communicating detailed, high-frequency information to an external controller. Consequently, such external computers would only be required to handle longer-term decisions using data inputs aggregated from vast multitudes (up to billions/trillions) of microbots. These microbots would also communicate with one another to compare the signals from nearby neurons, thereby providing a local context for signal interpretation. This localized processing will provide lower-latency interactions with neurons than are possible when the processing takes place far from neurons. This is required because the devices that collect neural signals (such as simple electrodes) lack signal-processing capabilities. The exploitation of this possibility requires the development of suitable control software (Freitas 2009). The limited volume and power of these microbots mean that they will likely contain diminutive, low-power computers, even if this translates to somewhat reduced computational speed. One theoretical possibility is mechanical molecular computers (Merkle *et al* 2018).

An important issue for BCI microbots will involve how to introduce these devices into the brain. Macroscopic electrodes and their connecting electrical leads require invasive surgery, while microscopic robots offer a much less invasive alternative via access through the circulation. Specifically, the microbots considered here would be small enough to traverse the capillaries (the smallest being ~3 μm in diameter), allowing them to reach within a few tens of microns of every cell in the brain via the bloodstream. The microbots could operate from within capillaries (e.g. by extending probes through capillary walls to access nearby cells). Alternatively, they might egress the capillaries via nanomedical diapedesis to their final destinations in proximity to target cells in the brain. Operating from within blood vessels is the simpler approach, but this limits the robots' ability to interact with individual neurons. Conversely, exiting the blood vessels to reach neurons will require more extensive safety evaluations due to the need to harmlessly pass through the blood–brain barrier (BBB) (Rustenhoven and Kipnis 2019). Safety concerns emphasize the design of microbots that minimally disrupt nearby tissues, even if this somewhat reduces their functionality or increases the complexity of their design or manufacture. The deployment of billions of such devices within the brain will allow for simultaneous access to many neurons in parallel, extending the precision and numbers of larger implanted devices, if required (Seo *et al* 2016). Whether operating from within capillaries or directly abutting neurons, microbots implementing BCIs will operate in microscopic environments in or near capillaries. This will require the evaluation of robot performance in such environments.

As context for this discussion, figure 2.1 is a schematic illustration of a capillary network that microbots could use to get close to cells in the brain. Inlet arteries and outlet veins are about a millimeter apart and correspond to typical capillary lengths. The capillary network consists of a set of short vessel segments joined to each other, as well as to the ends of the arterial and venous trees. The diameters of capillary vessels range from ~3–8 μm in diameter (Singhal *et al* 1973, Mühlfeld *et al* 2010).

Typical contents for the volume of capillary vessels in the diagram, based on standard values in the blood (adjusted for the reduced hematocrit), include ~3000 red blood cells, five white blood cells, and 100 microbots based on a total of ~10^{12} in circulation. The development of nanomedical robots involves two major challenges, the first being their fabrication in the potentially immense numbers required for medical applications. For example, depending on how comprehensively neurons are

Figure 2.1. Schematic of a capillary network (black) connecting the smallest branches of the arterial (red, at left) and venous (blue, at right) networks. The scale bar indicates the typical length scale for capillaries.

to be monitored, an advanced nanomedically enabled B/CI could require populations of such entities comparable to those of neurons and synapses within the brain (Martins *et al* 2019). To quantify this requirement, bacteria-sized robots would each have a mass of a few picograms. Thus, hypothetically, a trillion robots would involve (in aggregate) a few grams of devices introduced into the brain.

One approach for the fabrication of microscopic robots with nanometric components might be through the use of biological systems, e.g. RNA-based logic inside cells (Win and Smolke 2008), bacteria attached to nanoparticles (Martel *et al* 2008), the execution of simple programs via the genetic machinery within bacteria (Ferber 2004, Andrianantoandro *et al* 2006), DNA computers responding to specific chemical patterns (Benenson *et al* 2004), artificial DNA-based devices capable of self-locomotion (Smith 2010), or nucleic-acid-based atomic-scale manufacturing (Chen *et al* 2023). While these are promising initial steps, these systems lack the significant material, sensing, actuating, and computation capacities of the advanced microbots considered here (Freitas 1999, Nelson *et al* 2010, Boehm 2013), which will require significant improvements to current manufacturing technologies.

The second challenge will be to identify operational protocols that are appropriate for the physical environment of microbots. At these dimensions, viscous forces and Brownian motion are significant (Dusenbery 2009). Thus, the physics of microfluidics (Squires and Quake 2005) require different techniques than those used for larger robots (Purcell 1977). Moreover, the microbots must function reliably and safely in complex and highly variable biological milieus. This chapter considers this challenge by discussing how microbots might obtain power, move through tissue, communicate with other microbots, and navigate through microenvironments, particularly in small blood vessels.

2.2 Chemical power

Developing reliable power sources for medical implants is a significant challenge (Bazaka and Jacob 2013, Amar *et al* 2015). BCIs enabled by microbots will need to operate throughout the brain and for extended time periods to achieve their full potential. These requirements determine the best way to power microbots for BCIs.

Microscopic robots in the body could obtain power from a variety of sources (Freitas 1999, Nelson *et al* 2010), which form two broad categories. The first category involves externally delivered power, such as acoustic power from transducers on the skin. Such power sources are limited by the need to pass through the skull and encounter additional attenuation when reaching microbots deep within the brain. Moreover, external transducers are inconvenient for long-term use. The second category is power directly available within the body (e.g. from implanted batteries or reactive chemicals in the blood).

Fuel cells can power a variety of implanted medical devices (Barton *et al* 2004, Davis and Higson 2007), including BCIs (Rapoport *et al* 2012). Among these options, fuel cells that oxidize glucose are an appealing approach (Bazaka and Jacob 2013), as glucose and oxygen are available throughout the brain. These characteristics are especially useful for long-term monitoring, where it may not be convenient or feasible to deliver power from external devices, such as power from outside the body or implanted batteries. A variety of biological and artificial fuel cells can oxidize glucose (Chaudhuri and Lovley 2003), either completely to CO_2 or for several steps into the complete reaction. There are a variety of approaches toward the development of glucose fuel cells (Gogová *et al* 2010, An *et al* 2011, Zebda *et al* 2013), albeit the fabrication of efficient, reliable, and stable cells for the full oxidation of glucose remains a challenge.

The overall reaction that combines glucose and oxygen to produce water and carbon dioxide is:

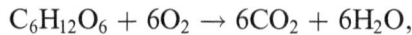

$$C_6H_{12}O_6 + 6O_2 \rightarrow 6CO_2 + 6H_2O,$$

which produces 4×10^{-18} J (Hogg and Freitas 2010). A fuel cell capable of the complete oxidation of glucose might capture about half this energy for use by a microbot. Because the concentration of glucose in the blood is significantly larger than that of oxygen (Freitas 1999), oxygen is the power-limiting chemical. Thus, an important measure is the reaction energy per oxygen molecule, which is one-sixth of the energy per glucose molecule.

Most of the oxygen carried by the blood is bound to hemoglobin via red blood cells. Microbots in the blood might consume oxygen from plasma (due to diffusion) as soon as it reached the plasma surface. The resulting reduced concentration in the plasma would draw oxygen from nearby cells. The kinetics of oxygen dissociation from hemoglobin would determine how rapidly cells could replenish oxygen in the plasma, with the time constant of this process being <100 ms (Clark *et al* 1985). Hence, for isolated microbots on timescales of a second or so, a reasonable approximation is that, as they traversed capillaries, the oxygen bound in cells would remain in equilibrium with the concentration in the plasma near the microbots, providing a relation between oxygen concentrations in the plasma and the oxygen bound within cells (Popel 1989). If, instead, dozens or hundreds of microbots happened to aggregate, they could conceivably extract oxygen from the plasma more rapidly than passing red blood cells could replenish it (Hogg and Freitas 2010).

A critical safety issue when microbots obtain power from oxygen is the extent to which this activity reduces the oxygen available for nearby tissues. For isolated robots (e.g. separated by at least tens of microns), the risk of oxygen depletion remains negligible. This is because microbot oxygen consumption is limited by the rate at which oxygen diffuses to the microbot's surface; thus, a single microbot cannot significantly deplete oxygen at typical concentrations in, or in proximity to, small blood vessels. However, oxygen depletion may be a significant issue for large populations of microbots or for large concentrations of robots in small, localized areas (e.g. if they are coordinating activities to interact with nerves within a small volume).

For the evaluation of microbot behavior, a convenient measure of chemical concentrations in a fluid is the number of molecules per unit volume, as this relates directly to units that are commonly used for molecular-scale events (e.g. diffusion) that are relevant for microscopic machines. Units convenient for larger scales include moles of a chemical per liter of fluid (i.e. molar units, M) and grams of a chemical per cubic centimeter. Furthermore, discussions of gases dissolved in the blood often specify concentrations indirectly via the corresponding partial pressure of the gas under standard conditions. As an example, an oxygen concentration of $C_{O2} = 10^{22}$ molecules m^{-3} corresponds to a 17 μM solution, 0.53 μg cm^{-3}, or a partial pressure of 1600 Pa or 12 mm Hg. It is also 0.037cm^{3}O$_2$/100 cm^3 of tissue for an oxygen volume measured at standard temperature and pressure.

For instance, figure 2.2 shows a group of microbots stationed around the wall of a small blood vessel. This figure corresponds to a small portion of a vessel and surrounding tissue of the network shown in figure 2.1. These microbots could consume oxygen diffusing to their surfaces to produce power. This consumption would alter the distribution of oxygen in the vessel and nearby tissues, as shown in figure 2.3 for a situation with moderately high oxygen demand by the tissue (Hogg and Freitas 2010). For this example, the microbots consume all the oxygen that reaches their surfaces, i.e. microbot power is diffusion limited.

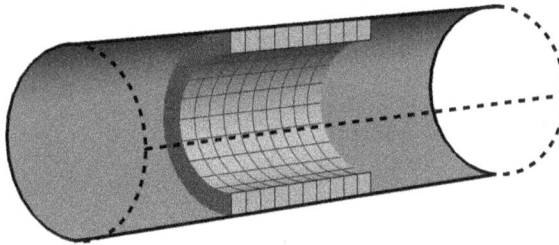

Figure 2.2. Schematic of a group of microbots (gray boxes, each 1 μm in length) aggregated around the wall of a small vessel. Half of the microbots and the vessel wall are cut away for clarity (indicated by dashed curves).

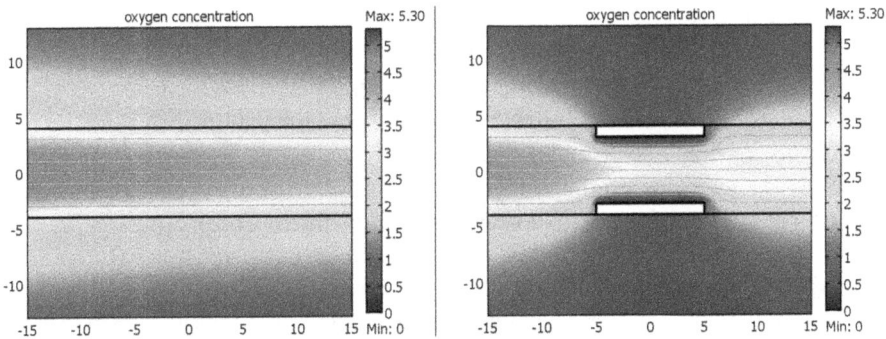

Figure 2.3. Oxygen concentrations in tissue and plasma within a blood vessel. Each plot shows a cross section through the vessel and surrounding tissue 30 μm in length. Typically, this length of vessel contains about four red blood cells. The left plot shows the vessel without microbots. The right plot includes a 10 micron ringset (the components that enable microbot operation and interactions with its environment) with pumps, which occupies the circumferential volume indicated by the white rectangles next to the vessel wall, which is a cross section through the microbots shown in figure 2.2. Fluid in the vessel flows from left to right. Distances along the sides of each plot are indicated in microns, and the concentrations shown on the color bars are in units of 10^{22} molecules m^{-3}. The horizontal black lines are the vessel walls, and the gray curves inside the vessel are fluid flow streamlines.

The microbots reduce the local oxygen concentration to a far greater degree than its consumption by the surrounding tissue, which can be seen by comparing vessels with and without microbots. Most of the extra oxygen used by the robots derives from the passing blood cells, which contain ~100 times the oxygen concentration of the plasma. Within the vessel containing the microbots, the oxygen concentration in the plasma is lowest in the fluid next to the microbots. Downstream of the microbots is a recovery region where the concentration increases slightly, as cells respond to the abruptly lowered oxygen concentration near the microbots. The streamlines in figure 2.3 show that the laminar flow speeds up as the fluid passes through the narrower vessel section where the microbots are stationed.

Studies such as those illustrated in figure 2.3 indicate that glucose oxidation could provide isolated robots with hundreds of picowatts of power, while aggregated robots could obtain tens of picowatts (Hogg and Freitas 2010). This is akin to the typical power consumption of human cells (Freitas 1999, Milo and Phillips 2015).

This discussion illustrates the average power that is available to microbots, which are relevant to the long-term operation required for the seamless functionality of a brain–machine interface. Microbots could also be endowed with a means of storing energy, thereby allowing them to utilize much higher power for short bursts and also to maintain operation during temporary anaerobic conditions, such as when white blood cells move through capillaries.

2.3 Locomotion

The bloodstream provides a convenient pathway for micron-sized microbots to access the brain and traverse the BBB, since they are small enough to pass through capillaries. Microbots in the circulation can move passively with the prevailing flow; however, this will deliver fewer microbots to sites with less-than-average blood flows. For BMIs, robots might employ active positioning, whether simply moving the short distance to the wall of a capillary in the brain, selecting among the bifurcations in a capillary network (figure 2.1), or egressing capillaries to reach individual neurons. These capacities will require that microbots move independently of the blood flow, at least for short distances.

Propulsion strategies appropriate for microscopic robots must account for the dominance of viscous drag, which necessitates novel forms of propulsion in contrast to the familiar methods employed for larger organisms and machines (Purcell 1977). A variety of approaches are possible (Freitas 1999, Nelson *et al* 2010). One example involves propulsion by magnetic fields (Dreyfus *et al* 2005, Abbott *et al* 2009), which can move ferromagnetic nanoparticles containing microbots through blood vessels (Ishiyama *et al* 2002, Martel *et al* 2007, Olamaei *et al* 2010).

Microorganisms utilize variously evolved locomotion mechanisms, which typically involve the actuation of extended structures, such as flagella and cilia (Jahn and Votta 1972, Purcell 1977, Brennen and Winet 1977, Lauga and Powers 2009, Guasto *et al* 2012). Demonstrated state-of-the-art micromachines suggest that biomimetic appendages akin to flagella (Behkam and Sitti 2007, Martel *et al* 2008, Zhang *et al* 2009, Qiu *et al* 2014) or cilia (Zhou and Liu 2008) may be a viable means of microbot propulsion.

However, such accoutrements present significant production and operational challenges for *in vivo* operations in proximity to cells and other microbots. The challenges involved in the fabrication of synthetic flagellar or ciliary appendages include their individual assembly (ideally self-assembly) and attachment to microbot surfaces. A significant operational challenge will be the potential for damaging nearby cells and/or tangling with nearby microbots. Moreover, these appendages would expose large surface areas to the biological milieu, leading to biofouling or immune reactions during extended use *in vivo*. The reliable avoidance of these issues will significantly increase the complexity of the microbot controller. Should a microbot need to shut down (e.g. due to component failure), its appendages might become entangled by subsequent passive motion in fluids or due to the movement of nearby cells. These issues will be relevant when deploying multitudes of microscopic robots in the brain. They might be imbued with collision-avoidance capacities to avoid harming any of the myriad nerve cell extensions.

A propulsion technique that avoids the difficulties associated with long external structures is via the motion of the microbot surface itself without the requirement for extended appendages. While not as commonly investigated as propulsion via flagella or cilia, some species of microorganisms achieve motility without the use of appendages (Ehlers *et al* 1996, Ehlers and Koiller 2011, Leshansky *et al* 2007). Two such strategies include steady undulating motions of surfaces flush with the microbot surface and small-amplitude surface oscillations (Hogg 2014).

Periodic surface oscillations can propel a microbot via diminutive parallel traveling surface waves, in either the opposite direction or the same direction as that of the wave motion (Brennen 1974). These waves traverse the full extent of the surface (e.g. from the north pole to the south pole of a sphere), even though the individual points on the surface are displaced by only a small distance. Figure 2.4 provides an example of fluid behavior at one instant of a traveling wave designed to deliver propulsion for a spherical entity (Blake 1971). The wave amplitude is small near the top and bottom of the sphere and large near the equator. The magnitudes of fluid velocities and pressure variations decrease rapidly away from the oscillating surface. The stresses induced in the fluid by the microbot are relatively small and decrease rapidly with distance from the robot surface; thus, the propulsive forces emanating from microbots passing nearby cells are unlikely to damage them (Hogg 2014).

Although microbot locomotion using surface waves is not as efficient as other methods (e.g. extended structures), it may be adequate for the fine positioning of microbots involved in BMIs, where microbots need only move over distances of tens of microns within or near small blood vessels to reach their target locations and do not need to move much further once stationed at these sites. For instance, waves with maximum amplitudes of 50 nm can move a micron-sized robot at 100 μm s^{-1} in fluids with viscosity similar to that of water, while dissipating significantly less than 1 pW (Hogg 2014). Similar waves and powers translate to a speed of 1 μm s^{-1} in fluids that are 10^4 times more viscous than water, which is a typical value for mucus or cell cytoplasm (Freitas 1999). At such speeds, microbots could access locations several tens of microns from capillaries, corresponding to a few cell diameters, in less

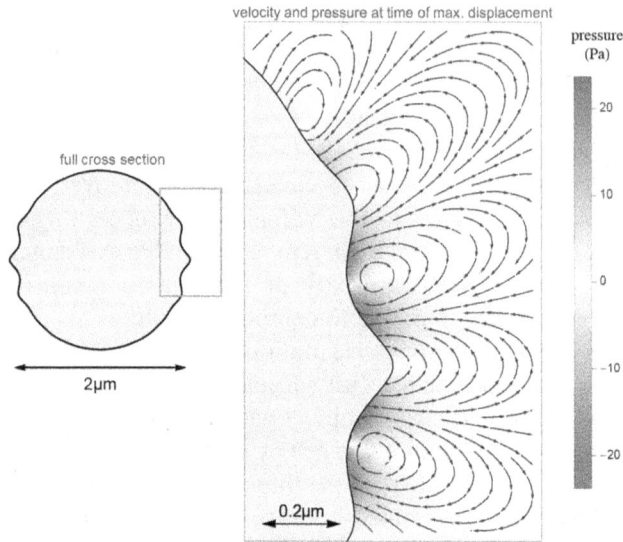

Figure 2.4. Velocity streamlines and pressures in a fluid in proximity to an oscillating sphere during an oscillation event of maximum displacement. Surface sites with positive pressure (green) move outward, while sites with negative pressure move inward. The rectangular box on the full cross section at the left corresponds to the region shown in detail on the right.

than a minute. Since all cells reside within a few cell diameters of capillaries, this locomotion approach could position the microbots involved in a selected brain interfacing platform using a small fraction of the available power, as discussed in section 2.2.

2.4 Communications

Microbots might employ a variety of methods to communicate, both with other nearby robots and to exchange data with external transceivers to integrate, for instance, a BMO (a robot that can engage with users through voice interactions) with external computers (Freitas 1999). Long-range communications could utilize techniques akin to those of larger-scale implants, including transducers that are affixed to the surface of the body or macroscopic implants. Microbots will face additional challenges, albeit opportunities, to exploit communications to coordinate the activities of neighboring microbots as they interact with different portions of a cell or nearby cells that directly influence each other (e.g. neurons connected at synapses). This would involve communications over tens to hundreds of microns.

Such short-range communications will enable groups of nearby microbots to arrive at decisions based not only on their own sensor measurements but also on those of their neighbors. Such comparisons will allow microbots to, for instance, move to optimal positions to best cover a region of tissue, transmit messages from deep within the brain to robots closer to external receivers, and eliminate redundant data to reduce the communication rates required for long-distance transmissions. Communications between neighboring robots may also facilitate their capacity to merge to create larger cooperative structures, which could increase the power and therapeutic capacities available to act on nearby tissues, both for signaling and (if necessary) the repair of damaged nerves (Hogg and Sretavan 2005).

This section discusses one such method, acoustic communication, which is suitable for use over distances of ~100 μm (Hogg and Freitas 2012). We consider both isolated microbots and groups thereof, up to about 10 μm in size. Such groups include microbots aggregated within blood vessels, giving a larger effective radiative size and the ability to direct acoustic waves, as the size of these aggregates would be comparable to the wavelengths of useful communication frequencies. Ultrasonic communications will not interfere with the electrical signals the microbots might use to measure or influence neural behavior.

Important acoustic properties for communications are the speed of sound, acoustic impedance between tissues, and attenuation among tissues, where the speed of sound determines communication latency. For microbots operating in the brain, short-range communications will involve transmission through tissues with homogeneous acoustic properties, with the speed of sound in soft tissues being ~1500 ms^{-1} (Freitas 1999). Although the boundaries between different types of tissues reflect acoustic waves, it is reasonable to ignore reflections from tissue boundaries at this scale. The reason for this is that the amplitudes of these reflections are contingent on the difference between the acoustic impedances on either side of the boundary, which is negligible because the impedances of most tissues cluster narrowly between 1.4 and 1.8 × 10^6 kg (m^{-2} s) (Freitas 1999). Scattering due to tissue heterogeneities is small because the wavelength at 100 MHz in water is 15 μm, which is much larger than the typical 10–500 nm dimensions of intracellular organelles and other potential scattering foci in tissue.

The skull is acoustically quite distinct from tissues, which limits the effectiveness of acoustic transmission to or from external sources. Specifically, the acoustic impedance of the skull is considerably larger, at ~8 × 10^6 kg (m^{-2} s^{-1}) (Freitas 1999), leading to a significant reflection of sound that reaches the skull through tissues. This is a critical factor for acoustic communications with sources external to the body, but not for those between nearby microbots in the brain.

Attenuation varies considerably with frequency, while for pure fluids such as water, attenuation increases quadratically with frequency; in other words, it increases by a factor of 100 over the frequency range from 10 to 100 MHz. At frequencies up to ~10 MHz, biological tissues also exhibit a power-law increase in attenuation with frequency, albeit with an exponent mainly in the range from one to 1.5. However, at the higher frequencies relevant for robot communications, this exponent gradually increases toward that of water. For the brain, the exponent is close to one; thus, attenuation at frequency f is $\alpha = af$ where $a = 10^{-5}$ s m^{-1} (Freitas 1999). Therefore, for instance, at 10 MHz, the attenuation length scale $1/\alpha$ is about a centimeter. Thus, the decrease in acoustic power due to attenuation, which is proportional to $\exp(-2\alpha d)$ over a distance d, is small for communications over tens of microns for microbots coordinating their activities with neighboring cells. Instead, the main reduction in power arises from the propagation of acoustic energy, which is proportional to $1/d^2$. This spread is particularly significant, as the microbots are too small compared to the wavelength of sound to form directed beams. Conversely, groups of neighboring microbots that can coordinate their oscillations could form directed beams with less spreading (Hogg and Freitas 2012).

Figure 2.5 shows an example of the acoustic behavior near a ringset with oscillations on its outer surface, where the narrow vertical black lines indicate the locations of the vessel walls. Unlike a spherical oscillator, the oscillation of the ringset surface conveys significant directionality to the beam at higher frequencies. As an estimate of communication distances for neighboring microbots, for a billion microbots distributed throughout the brain, which has a volume of about one liter, the typical distance between neighbors would be $\sim$100 μm.

For communication distances of over $\sim$100 μm in tissue such as the brain, frequencies in the tens of megahertz provide the best compromise between acoustic efficiency (better at higher frequencies, with wavelengths short enough to be comparable to the size of the microbots) and attenuation (better at lower frequencies). In this frequency range, acoustic wavelengths are tens of microns, an order of magnitude larger than the microbots. This range of frequencies can provide communication rates of $\sim$10^4 bits s^{-1} using 100 pW of broadcast power. The resulting power flux, even near the transmitting robot, is within accepted safety limits for medical ultrasound (Hogg and Freitas 2012), though care will be required when groups of nearby robots communicate simultaneously to avoid regions of higher power due to constructive interference.

Communications could extend over larger distances with burst power, as discussed in section 2.2, via message-passing networks of microbots or the grouping of some microbots into larger aggregates that can efficiently transmit at lower frequencies where tissue attenuation is lower (Freitas 1999).

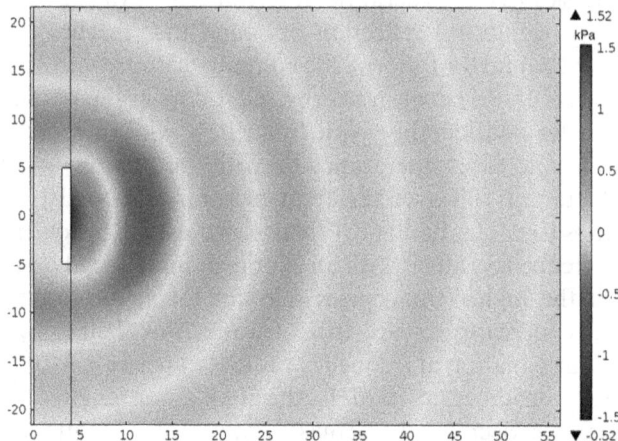

Figure 2.5. Pressure variations near a microbot ringset oscillating at 100 MHz at times of peak pressure. The calculation assumes low-attenuation tissue. The distances along axes are in microns, and the pressure is in kPa. The vertical line indicates the vessel wall, and the white rectangle denotes the microbots next to the vessel wall. The figure is a cross section through an axially symmetric vessel and tissue domain.

2.5 Navigation

Creating BMOs, BCIs, or B/CIs using microscopic robots will require microbots to be guided to myriad specific sites within the brain. This navigation will encompass a range of scales, including coarse navigation to initially guide microbots to the brain, distribution through capillary networks within the brain, and finally fine navigation over tens of microns to the exact locations required for interacting with specific neurons, either on the blood vessel walls or directly next to neurons.

A variety of techniques can provide navigation at these body, organ, and cell-length scales (Freitas 1999, Nelson *et al* 2010). For instance, external signals could provide coarse localization for the robots. For microscale navigation, onboard sensors that detect chemical, thermal, and other properties could allow robots to identify locations within organs based on their distinctive attributes.

An example of small-scale navigation is microbots traversing capillaries, as illustrated in figure 2.1. The trajectories microbots use will depend on how they select between the available routes at the bifurcations they encounter. For instance, a BCI may require robots to be uniformly distributed throughout the brain or concentrated in specific regions contingent on neural activities. The desired robot distribution may depend on the global properties of a network, such as the number of metabolically active cells around each segment in a capillary network. In some cases, measurable physiological properties at early branch points in a network could indicate where the robots should go (e.g. if there is more blood flow toward active regions, the microbots could follow that flow), if such indications are not sufficient between the blood vessel bifurcations.

The simplest approach may entail random selection among branches and continued circulation until robots identify a suitable location that is not already occupied by an adequate population of microbots. With a sufficiently large population of microbots and enough time, random choices could guide them to the desired locations, and this process could be expedited through their cooperation. For example, when approaching a bifurcation, a microbot could pick the one not yet visited by other microbots if earlier visiting microbots leave markers to indicate their choices; an example of coordination using stigmergy.

Alternatively, for branch choices made close to the final locations, microbots that discover a suitable location could signal to other robots to come to or avoid that region, as appropriate (Hogg 2007), by employing short-range communications, as discussed in section 2.4. Communication between neighboring microbots or leaving markers on vessel walls could enable this cooperative exploration of the network as a more efficient approach than random choices.

Microscopic robots may not have sufficient computational and communication capacities to accurately implement such cooperative protocols. Nevertheless, the vast number of microbots may, with a high probability, ensure that at least some of them reach the desired locations, even if they use simpler protocols. For instance, suppose microbots make branch choices independently using only locally available information. These data might include flow speed, branching geometries, and chemical concentrations; since velocities and concentrations could vary over time, this would lead to time-dependent branch choices. A further simplification would occur when the local information does not change significantly over the timescale required to properly distribute the microbots to desired sites within the brain.

The data available to a microbot that reaches a bifurcation may depend on how it arrives (e.g. near the vessel wall or near its center), which in turn could depend on how it entered a specific vessel segment at an earlier branch point. The position of a microbot within a vessel will also affect how it moves passively with the blood flow. In general, the ability of a microbot to assess data could rely on its travel history. For example, if it happens to follow a particularly long and slow path to attain a branch point, it may have inadequate power for the evaluation of sensor measurements. If such a dependence on history is not important, two simple navigation possibilities are: (1) microbots move passively with the flow such that they select a branch that is proportional to the fluid flow rate into each branch, and (2) microbots pick a branch uniformly at random.

Figure 2.6 illustrates how these navigation choices may affect how the microbots distribute themselves through a capillary network. In this exemplar geometry, flow-based movement distributes microrobots more evenly across capillary segments than uniform branching, as the latter has relatively fewer microbots passing through segments that are accessible only after several bifurcations.

On the other hand, uniform branch divisions would guide more microbots to segments with low flow that are only a few branchings from the arterial inlet. This example uses a simplified model of flow, namely two-dimensional steady-state flow with uniform fluid properties. These simplifications give an approximately averaged behavior, which is suitable for estimating how a large population of transiting robots would be distributed through the vessels. This example does not include subtle flow variations, as cells move and occasionally block vessel segments, particularly large white blood cells. Nor does this steady-state behavior account for temporal variations in the flow due to changes in the blood flow distribution at macroscopic scales or when capillaries are open (Feher 2017).

Microbots dispersing through capillaries according to these or other schemes will require active navigation and locomotion. This would be true even for microbots

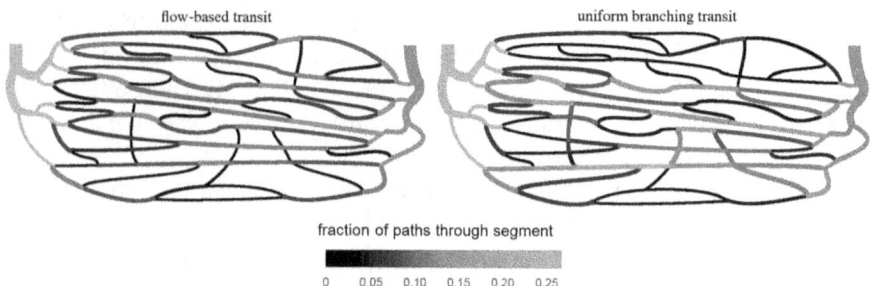

Figure 2.6. Fractions of pathways through each segment of the capillary network in figure 2.1 for two microbot navigation methods. Lighter colors and thicker lines indicate a larger fraction. The arterial and venous trees on the left and right sides of the network are shown for context. The two cases illustrated here are for microbots following the blood flow (left) and those selecting each branch with equal probability (right).

that follow the overall flow, since they will need to compensate for branching effects that selectively emphasize plasma or objects, such as cells or microbots, i.e. plasma skimming and the Fahraeus effect (Pries *et al* 1996).

A computationally simple navigation strategy based on data that is locally available to microbots can employ the patterns of stresses on their surfaces. This is analogous to the navigation used by some fish (Bleckmann and Zelick 2009), albeit with a significantly different physical regime for flow: a laminar, viscosity-dominated regime for microscopic robots, rather than the turbulent, inertia-dominated flows encountered by fish. Specifically, a fish can orient itself and detect nearby objects based on how the pressure patterns on its surface change as it moves. In general fluid flows, the interpretation of stresses is a computationally demanding task (Bouffanais *et al* 2011); although, in some cases, machine learning can generate models that are easier to evaluate once trained (Raissi *et al* 2019). Fortunately, smooth microscopic flows give a linear relation between stress and flow speed, which greatly simplifies the computations required to interpret surface stress patterns (Hogg 2018). In addition, navigation via stresses is a passive measurement, which avoids any safety issues that might arise from actively probing the cells around microbots (e.g. acoustically). Microbots selecting between bifurcations in capillary networks or operating outside capillaries will require active locomotion (section 2.3), which will also produce stresses on the microbot surface based on nearby objects.

As an example of the data that might be available to a microbot due to stresses, figure 2.7 represents the stress patterns on its surface as it is being pushed by fluid

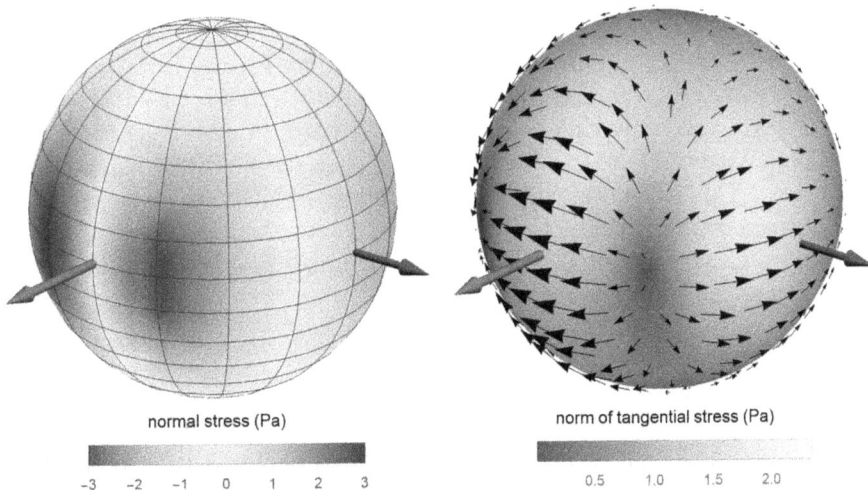

normal stress (Pa)   norm of tangential stress (Pa)

Figure 2.7. Exemplar pattern of stresses on a microscopic sphere moving with the fluid flow near a capillary wall. The blue arrows on the right of each sphere indicate the direction of motion. The gray arrows indicate the direction to the nearest point on the vessel wall. (a) Normal stress: positive and negative values correspond to tension and compression, respectively. (b) Tangential stress: arrows show the direction and relative size of the tangential stress vector, and the color indicates its norm. Hogg (2018) gives the parameters of the geometry and flow for this example.

through a vessel that is comparable in size to a capillary. The figure decomposes the stress into components that are: (i) normal and (ii) tangential to the microbot surface. The stress is relatively large on the side of a sphere that is facing a nearby vessel wall.

Furthermore, the most significant variation in the stress is in the azimuthal direction around the sphere's equator. In particular, the largest magnitude of tangential stress is at the point of the sphere that is in closest proximity to the vessel wall. The direction of motion of the sphere is $\sim 90°$ around the equator from that closest point. These quantified observations will allow the microbot to employ surface stresses to estimate its orientation, distance from the vessel wall, and motion within the vessel itself (Hogg 2018). Using these data, microbots can often determine when they encounter bifurcating vessels (Hogg 2020). Stress-based navigation is well suited to micron-sized robots. This is because they are both small enough to traverse capillaries and sufficiently large to have a useful signal-to-noise ratio when averaged over the time a microbot transits a distance comparable to its own size. Neither larger robots nor significantly smaller devices, such as nanoparticles, satisfy both conditions.

The quantification of surface stresses may facilitate a variety of navigation applications. For example, decreasing estimates of a vessel diameter indicate a narrowing vessel where the microbot may become trapped. By responding to this information, the microbot might avoid damage to itself or its surroundings.

Orientation estimates would allow a microbot to determine its upstream and downstream sides, which could improve chemical gradient measurements (Dusenbery 2009). Knowing its orientation with respect to a vessel wall could enable a microbot to move directly toward the vessel wall, thereby reaching it more rapidly and using less energy than that used in a blind search. This could assist microbots with the formation of aggregate structures (Freitas 1999) or locate the sources of chemicals released from the vessel wall (Hogg and Kuekes 2006). Furthermore, orientation estimates could help control microbot locomotion (e.g. to maintain a fixed orientation with respect to vessels despite Brownian motion and fluid torques). This might simplify a subsumption architecture (Brooks 1991) for microscopic robots, since higher-level controls operating at longer timescales would not then need to account for these unintended changes in orientation.

2.6 Discussion

This chapter is a theoretical study of some capabilities of microbots that cannot yet be manufactured. This means that the microbots discussed here cannot yet be tested, tuned, or optimized for safe and efficacious use to enable BMO, BCI, and a potential future B/CI. Instead, quantitative models must be relied upon to estimate their potential for these applications, as well as their limitations. As described in this chapter, such models indicate that microbots could obtain enough power to support the locomotion, communication, and navigation required to operate various brain–interface platforms. Thus, microbots are an appealing potential technology for the realization of these platforms.

The capabilities discussed here demonstrate various design trade-offs for microbots due to their limited volumes, surface areas, and power. These constraints motivate the use of components for multiple purposes. Specifically, actuated surfaces with force sensors can provide traveling waves for locomotion at kilohertz frequencies (section 2.3), megahertz communications (section 2.4), and stress-based navigation by measuring variations at ~100 hertz (section 2.5). These are not necessarily the most effective strategies for each of these capabilities when considered individually; however, shared hardware might be more easily housed within the limited volumes and surface areas of microbots than distinct hardware platforms for each capability.

Vibrating surfaces can also provide mechanical power from ultrasound (Freitas 1999) as an alternative to the chemical power discussed here. Finally, suitable oscillations can temporarily open the BBB to facilitate drug delivery to the brain (Underwood 2015). For example, experiments in mice identified the range of ultrasound pressure variations in small vessels that can temporarily open the BBB without harm to the tissue (Pascal-Tenorio *et al* 2018, Ogawa *et al* 2022). Alternatively, hyperthermia induced by magnetic nanoparticles can briefly open the BBB, enabling drug delivery to the brain (Boehm and Chen 2009, Tabatabaei *et al* 2015, Gupta *et al* 2022). This may also assist microscopic robots with egressing from capillaries in the brain using locomotion based on the surface motions discussed in section 2.3.

The discussion in this chapter has focused on individual, well-distinguished microbots or small numbers thereof operating in proximity. BMO, BCI, and a hypothetical B/CI might utilize large populations of robots throughout the brain. Thus, a critical question is whether and how these microbots might affect each other and the overall physiology of the brain. For instance, individual microbots using oxygen for power would have a negligible effect on the amount of oxygen that is typically available to tissues. However, large numbers of such microbots might have significant detrimental impacts, particularly in view of the high metabolic demands of neurons. Similarly, large numbers of communicating robots could interfere with each other and necessitate stringent control protocols, thereby reducing the computation available for their primary task of interpreting neural signals in support of brain/machine, computer, or cloud (edge) interface applications. In the physiological domain, the energy that microbots dissipate during their operation would radiate heat into their immediate environment. Dozens to hundreds of adjacent microbots that consume oxygen as fast as it reaches them would translate to negligible temperature increases (Hogg and Freitas 2010). However, if considerably larger populations of robots needed to operate simultaneously in relatively small regions of the brain, they would have to ensure that their dissipated heat did not damage nearby tissues.

Prior to the application of microbots to enable various types of brain interfaces, they will likely serve as useful research tools for *in vivo* measurements and interventions at cellular/organellar resolution. They may complement studies of fixed tissues, such as mapping the connectome (DeWeerdt 2019) and microvasculature (Cassot *et al* 2006, Boehm 2013), as well as macroscale studies of brain activities and manipulation (e.g. using implanted electrodes or magnetic fields) (Grossman 2018). These studies will assist with the formulation of protocols for the safe and effective introduction of microbots into the brain. Such experimental evaluations will quantify physiology at the micron scale within the heterogeneous environments and complex 3D structures of connections in the brain, as revealed by state-of-the-art imaging techniques (Motta *et al* 2019). Research microbots will also be able to refine our understanding of how memories are physically encoded in the brain (Josselyn and Tonegawa 2020), which will aid in the development of techniques for the use of future generations of microbots for fully operational BMIs, BCIs, and an envisaged B/CI. This information will guide the design of the microbots and their control software while also complementing the use of quantitative models of robot capabilities, such as those described in this chapter, and improving their application to a variety of brain interfaces.

Acknowledgments

I thank Robert Freitas Jr., Ralph Merkle, and James Ryley for helpful discussions.

References

Abbott J J, Peyer K E, Lagomarsino M C, Zhang L, Dong L, Kaliakatsos I K and Nelson B J 2009 How should microrobots swim? *Int. J. Robot. Res.* **28** 1434–47

An L, Zhao T S, Shen S Y, Wu Q X and Chen R 2011 Alkaline direct oxidation fuel cell with non-platinum catalysts capable of converting glucose to electricity at high power output *J. Power Sources* **196** 186–90

Andrianantoandro E, Basu S, Karig D K and Weiss R 2006 Synthetic biology: new engineering rules for an emerging discipline *Mol. Syst. Biol.* **2** 2006.0028

Barton S C, Gallaway J and Atanassov P 2004 Enzymatic biofuel cells for implantable and microscale devices *Chem. Rev.* **104** 4867–86

Bazaka K and Jacob M V 2013 Implantable devices: issues and challenges *Electronics* **2** 1–34

Behkam B and Sitti M 2007 Bacterial flagella-based propulsion and on/off motion control of microscale objects *Appl. Phys. Lett.* **90** 023902

Amar A, Kouki A B and Cao H 2015 Power approaches for implantable medical devices *Sensors (Basel)* **15** 28889–914

Benenson Y, Gil B, Ben-Dor U, Adar R and Shapiro E 2004 An autonomous molecular computer for logical control of gene expression *Nature* **429** 423–9

Betancourt T and Brannon-Peppas L 2006 Micro- and nanofabrication methods in nanotechnological medical and pharmaceutical devices *Int. J. Nanomedicine* **1** 483–95

Blake J 1971 A spherical envelope approach to ciliary propulsion *J. Fluid Mech.* **46** 199–208

Bleckmann H and Zelick R 2009 Lateral line system of fish *Integr. Zool.* **4** 13–25

Boehm F J (ed) 2013 *Nanomedical Device and Systems Design: Challenges, Possibilities, Visions* (Boca Raton, FL: CRC Press)

Boehm F J and Chen A 2009 Medical applications of hyperthermia based on magnetic nanoparticles *Recent Pat. Biomed. Eng.* **2** 110–20

Bouffanais R, Weymouth G D and Yue D K P 2011 Hydrodynamic object recognition using pressure sensing *Proc. R. Soc. A.* **467** 19–38

Brennen C 1974 An oscillating-boundary-layer theory for ciliary propulsion *J. Fluid Mech.* **65** 799–824

Brennen C and Winet H 1977 Fluid mechanics of propulsion by cilia and flagella *Annu. Rev. Fluid Mech.* **9** 339–98

Brooks R A 1991 New approaches to robotics *Science* **253** 1227–32

Cassot F, Lauwers F, Fouard C, Prohaska S and Lauwers-Cances V 2006 A novel three-dimensional computer-assisted method for a quantitative study of microvascular networks of the human cerebral cortex *Microcirculation* **13** 1–18

Chaudhuri S K and Lovley D R 2003 Electricity generation by direct oxidation of glucose in mediatorless microbial fuel cells *Nat. Biotechnol.* **21** 1229–32

Chen X, Yan B and Yao G 2023 Towards atom manufacturing with framework nucleic acids *Nanotechnology* **34** 172002

Clark A Jr., Federspiel W J, Clark P A and Cokelet G R 1985 Oxygen delivery from red cells *Biophys. J.* **47** 171–81

Collinger J L, Wodlinger B, Downey J E, Wang W, Tyler-Kabara E C, Weber D J *et al* 2013 High-performance neuroprosthetic control by an individual with tetraplegia *Lancet* **381** 557–64

Davis F and Higson S P 2007 Biofuel cells—recent advances and applications *Biosens. Bioelectron.* **22** 1224–35

DeWeerdt S 2019 How to map the brain *Nature* **571** S6–8

Dreyfus R, Baudry J, Roper M L, Fermigier M, Stone H A and Bibette J 2005 Microscopic artificial swimmers *Nature* **437** 862–5

Dusenbery D B 2009 *Living at Micro Scale: The Unexpected Physics of Being Small* (Cambridge, MA: Harvard University Press)

Ehlers K M, Samuel A D, Berg H C and Montgomery R 1996 Do cyanobacteria swim using traveling surface waves? *Proc. Natl. Acad. Sci. U. S. A.* **93** 8340–3

Ehlers K M and Koiller J 2011 Micro-swimming without flagella: propulsion by internal structures *Regul. Chaot. Dyn.* **16** 623–52

Feher J 2017 *Quantitative Human Physiology* 2nd edn (New York: Academic)

Ferber D 2004 Synthetic biology. Microbes made to order *Science* **303** 158–61

Freitas R A Jr. 1999 *Nanomedicine, Volume I: Basic Capabilities* (Georgetown, TX: Landes Bioscience)

Freitas R A Jr. 2009 Computational tasks in medical nanorobotics *Bio-Inspired and Nanoscale Integrated Computing* ed M M Eshaghian-Wilner (New York, NY: Wiley) 391–428

Gogová Z, Hanika J and Markoš J 2010 Optimal design of a multifunctional reactor for catalytic oxidation of glucose with fast catalyst deactivation *Dynamic Modelling* ed A V Brito (Rijeka, Croatia: InTech) 209–32

Grossman N 2018 Modulation without surgical intervention *Science* **361** 461–2

Guasto J S, Rusconi R and Stocker R 2012 Fluid mechanics of planktonic microorganisms *Annu. Rev. Fluid Mech.* **44** 373–400

Gupta R, Chauhan A, Kaur T, Kuanr B K and Sharma D 2022 Transmigration of magnetite nanoparticles across the blood-brain barrier in a rodent model: influence of external and alternating magnetic fields *Nanoscale* **14** 17589–606

Hogg T 2007 Coordinating microscopic robots in viscous fluids *Auton. Agents Multi-Agent Syst.* **14** 271–305

Hogg T 2014 Using surface-motions for locomotion of microscopic robots in viscous fluids *J. Micro-Bio. Robotics.* **9** 61–77

Hogg T 2018 Stress-based navigation for microscopic robots in viscous fluids *J. Micro-Bio. Robotics.* **15** 59–67

Hogg T 2020 Identifying vessel branching from fluid stresses on microscopic robots *Control Systems Design of Bio-Robotics and Bio-Mechatronics with Advanced Applications* ed A Taher Azar (Amsterdam: Elsevier) 171–200

Hogg T and Freitas R A Jr. 2010 Chemical power for microscopic robots in capillaries *Nanomed. Nanotechnol. Biol. Med.* **6** 298–317

Hogg T and Freitas R A Jr. 2012 Acoustic communication for medical nanobots *Nano Commun. Netw.* **3** 83–102

Hogg T and Kuekes P J 2006 Mobile microscopic sensors for high-resolution *in vivo* diagnostics *Nanomed. Nanotechnol. Biol. Med.* **2** 239–47

Hogg T and Sretavan D W 2005 Controlling tiny multi-scale robots for nerve repair *Proc. of the 20th Natl. Conf. on Artificial Intelligence (AAAI2005)* ed M Veloso and S Kambhampati (AAAI Press) 1286–91

Ishiyama K, Sendoh M and Arai KI 2002 Magnetic micromachines for medical applications *J. Magn. Magn. Mater.* **242–245** 1163–5

Jager E W, Inganäs O and Lundström I 2000 Microrobots for micrometer-size objects in aqueous media: potential tools for single-cell manipulation *Science* **288** 2335–8

Jahn T L and Votta J J 1972 Locomotion of protozoa *Annu. Rev. Fluid Mech.* **4** 93–116

Josselyn S A and Tonegawa S 2020 Memory engrams: recalling the past and imagining the future *Science* **367** eaaw4325

Lauga E and Powers T R 2009 The hydrodynamics of swimming microorganisms *Rep. Prog. Phys.* **72** 096601

Leshansky A M, Kenneth O, Gat O and Avron J E 2007 A frictionless microswimmer *New J. Phys.* **9** 145

Llinás R R, Walton K D, Nakao M, Hunter I and Anquetil P A 2005 Neuro-vascular central nervous recording/stimulating system: using nanotechnology probes *J. Nanopart. Res.* **7** 111–27

Martel S *et al* 2007 Automatic navigation of an untethered device in the artery of a living animal using a conventional clinical magnetic resonance imaging system *Appl. Phys. Lett.* **90** 114105

Martel S, Felfoul O and Mohammadi M 2008 Flagellated bacterial nanorobots for medical interventions in the human body *2nd IEEE RAS & EMBS Int. Robotics and Biomechatronics (Scottsdale, AZ)* 264–9

Martins N R B *et al* 2019 Human brain/cloud interface *Front. Neurosci.* **13** 112

Merkle R C, Freitas R A Jr., Hogg T, Moore T E, Moses M S and Ryley J 2018 Mechanical computing systems using only links and rotary joints *ASME. J. Mech. Rob.* **10** 061006

Milo R and Phillips R 2015 *Cell Biology by the Numbers* 1st edn (New York, NY: Garland Science)

Monroe D 2009 Micromedicine to the rescue *Commun. ACM* **52** 13–5

Morris K 2001 Macrodoctor, come meet the nanodoctors *Lancet* **357** 778

Motta A, Berning M, Boergens K M, Staffler B, Beining M, Loomba S, Hennig P, Wissler H and Helmstaedter M 2019 Dense connectomic reconstruction in layer 4 of the somatosensory cortex *Science* **366** eaay3134

Mühlfeld C, Weibel E R, Hahn U, Kummer W, Nyengaard J R and Ochs M 2010 Is length an appropriate estimator to characterize pulmonary alveolar capillaries? A critical evaluation in the human lung *Anat. Rec. (Hoboken)* **293** 1270–5

Nelson B J, Kaliakatsos I K and Abbott J J 2010 Microrobots for minimally invasive medicine *Annu. Rev. Biomed. Eng.* **12** 55–85

O'Doherty J E, Lebedev M A, Ifft P J, Zhuang K Z, Shokur S, Bleuler H and Nicolelis M A 2011 Active tactile exploration using a brain–machine–brain interface *Nature* **479** 228–31

Ogawa K *et al* 2022 Focused ultrasound/microbubbles-assisted BBB opening enhances LNP-mediated mRNA delivery to brain *J. Control. Release* **348** 34–41

Olamaei N, Cheriet F, Beaudoin G and Martel S 2010 MRI visualization of a single 15 μm navigable imaging agent and future microrobot *Annu. Int. Conf. IEEE Eng. Med. Biol. Soc.* **2010** 4355–8

Pascal-Tenorio A, Li N, Lechtenberg K J, Rosenberg J, Airan R D, James M L *et al* 2018 Acute activation of microglia and astrocytes after BBB opening with low intensity pulsed focused ultrasound and microbubbles in a mouse model *Proc. of 6th Intl. Symp. on Focused Ultrasound* 230–1

Popel A S 1989 Theory of oxygen transport to tissue *Crit. Rev. Biomed. Eng.* **17** 257–321

Pries A R, Secomb T W and Gaehtgens P 1996 Biophysical aspects of blood flow in the microvasculature *Cardiovasc. Res.* **32** 654–67

Purcell E M 1977 Life at low Reynolds number *Am. J. Phys.* **45** 3–11

Qiu F, Zhang L, Peyer K E, Casarosa M, Franco-Obregón A, Choi H and Nelson B J 2014 Noncytotoxic artificial bacterial flagella fabricated from biocompatible ORMOCOMP and iron coating *J. Mater. Chem. B.* **2** 357–62

Raissi M, Perdikaris P and Karniadakis G E 2019 Physics-informed neural networks: a deep learning framework for solving forward and inverse problems involving nonlinear partial differential equations *J. Comput. Phys.* **378** 686–707

Rapoport B I, Kedzierski J T and Sarpeshkar R 2012 A glucose fuel cell for implantable brain–machine interfaces *PLoS One* **7** e38436

Rustenhoven J and Kipnis J 2019 Bypassing the blood-brain barrier *Science* **366** 1448–9

Sánchez S and Pumera M 2009 Nanorobots: the ultimate wireless self-propelled sensing and actuating devices *Chem. Asian J.* **4** 1402–10

Seo D, Neely R M, Shen K, Singhal U, Alon E, Rabaey J M, Carmena J M and Maharbiz M M 2016 Wireless recording in the peripheral nervous system with ultrasonic neural dust *Neuron* **91** 529–39

Shih J J, Krusienski D J and Wolpaw J R 2012 Brain–computer interfaces in medicine *Mayo Clin. Proc* **87** 268–79

Singhal S, Henderson R, Horsfield K, Harding K and Cumming G 1973 Morphometry of the human pulmonary arterial tree *Circ. Res.* **33** 190–7

Smith L M 2010 Nanotechnology: molecular robots on the move *Nature* **465** 167–8

Squires T M and Quake S R 2005 Microfluidics: fluid physics at the nanoliter scale *Rev. Mod. Phys.* **77** 977–1026

Tabatabaei S N, Girouard H, Carret A S and Martel S 2015 Remote control of the permeability of the blood-brain barrier by magnetic heating of nanoparticles: a proof of concept for brain drug delivery *J. Control. Release* **206** 49–57

Thomas T P, Shukla R, Majoros I J, Myc A and Baker J R Jr. 2007 Polyamidoamine dendrimer-based multifunctional nanoparticles *Nanobiotechnology II: More Concepts and Applications* ed C Mirkin and C Niemeyer (Hoboken, NJ: Wiley-VCH Press) 305–19

Thubagere A J, Li W, Johnson R F, Chen Z, Doroudi S, Lee Y L *et al* 2017 A cargo-sorting DNA robot *Science* **357** eaan6558

Underwood E 2015 Neuroscience. Can sound open the brain for therapies? *Science* **347** 1186–7

Win M N and Smolke C D 2008 Higher-order cellular information processing with synthetic RNA devices *Science* **322** 456–60

Zebda A, Cosnier S, Alcaraz J P, Holzinger M, Le Goff A, Gondran C *et al* 2013 Single glucose biofuel cells implanted in rats power electronic devices *Sci. Rep.* **3** 1516

Zhang L, Abbott J A, Dong L, Kratochvil B E, Bell D and Nelson B J 2009 Artificial bacterial flagella: fabrication and magnetic control *Appl. Phys. Lett.* **94** 064107

Zhou Z-G and Liu Z-W 2008 Biomimetic cilia based on MEMS technology *J. Bionic Eng.* **5** 358–65

Chapter 3

Combining a neural bypass and a neural allograft to develop a prosthetic thalamus

Miguel Pais-Vieira, António J Salgado and Carla Pais-Vieira

A system and strategy are proposed here for the restoration of communication after the formation of lesions in the nervous system. Let us consider the three domains in the nervous system that sequentially transmit information as follows: the first region sends afferent (e.g. sensory) information to a second relay site (e.g. the thalamus), from which it is transferred to a third target region (e.g. the cortex). We propose a system and method consisting of two sequential implants: (a) first, a neural bypass that bridges the lesioned site and (b) second, a graft of neurons into the lesioned site. The neural bypass facilitates information transfer between the afferent and target regions, while the neural graft performs some of the original computations and promotes communication with adjacent structures. Consequently, the neural bypass ensures that the original functionality is restored, while the neural graft promotes the long-term recovery of the actual biocircuitry. This approach comprises six different phases: (1) neural bypass implant, (2) neural graft development, (3) neural graft training, (4) remote neural graft control, (5) neural allograft implant, and finally (6) bypass removal. Although the strategy proposed here would involve six phases, phases three and six may be adapted contingent on the responses of the subject to each of the implants, as well as the functional status of the original region.

Preamble
The present text delineates a series of steps to test the possibility of combining neural bypasses with neural allografts. This approach will require expertise from multiple fields of knowledge; thus, several concepts from distinct specialties may be employed. The organization of this text will begin by defining the main terms (Definitions), followed by the different phases that are expected to be followed. Throughout the text, the experiments required to test the underlying hypotheses related to each development (i.e. phase) of the system are described. Finally, it should be noted that multiple techniques proposed here for recording and stimulation can, in theory, be replaced by

doi:10.1088/978-0-7503-2144-0ch3
3-1

other techniques (e.g. electrical microstimulation, optogenetic stimulation), with the proviso that the changes resulting from these modifications are considered (e.g. prodromic and antidromic stimulation).

3.1 Introduction

The regeneration and/or restoration of functionality in various regions of the central nervous system remain challenging and complex goals for both research and clinical scientists (Horner and Gage 2000, Illis 2012, Teixeira *et al* 2013, Williams 2014, Lebedev and Nicolelis 2017, Jensen *et al* 2020, Varadarajan *et al* 2022). Multiple approaches have been undertaken to achieve advances in these areas. These strategies can range from brain–machine interfaces (Donati *et al* 2016, Shokur *et al* 2018, Selfslagh *et al* 2019, Pais-Vieira *et al* 2022, Rowald *et al* 2022), to mesenchymal stem cells (Teixeira *et al* 2013, Anbiyaiee *et al* 2020, de Araújo *et al* 2022), and other techniques (Lima *et al* 2006, Dlouhy *et al* 2014, Tsintou *et al* 2015, Marsh and Blurton-Jones 2017). Current techniques that facilitate the replacement, regeneration, or enhancement of the functionalities of the central nervous system typically require highly complex state-of-the-art procedures, with variable results (Cicchetti *et al* 2009, Dlouhy *et al* 2014, Donati *et al* 2016, Cisbani *et al* 2017, Shokur *et al* 2018, Selfslagh *et al* 2019, Pais-Vieira *et al* 2022, Rowald *et al* 2022). Here, we propose a roadmap for the development of a prosthetic thalamus through the combination of two different techniques: neural bypasses (Pais-Vieira *et al* 2013, Pais-Vieira *et al* 2015, Sharma *et al* 2016, Bouton 2019) and neural allografts, using biological scaffolds (Silva *et al* 2010, Xue *et al* 2021, Zhang *et al* 2019).

The general underlying idea for the present approach is that neural bypasses may be employed as a form of software that can lead to neural plasticity in the central nervous system, while neural allografts (spatially organized as columns of neurons) can be thought of as a type of hardware. This hardware can be trained using the information from the neural bypass. Over the short to medium term, the spatial organization guided by the biological scaffold that contains the neural allografts would benefit from the support of the geometrically organized neuronal columns, ensuring that the information is processed and transmitted via specific pathways. Meanwhile, the neural bypass would temporarily provide the software to ensure that specific functions are achieved. Over the medium to long term, the neural allograft would establish functional connections with the relevant areas (Hebb 2002). Furthermore, the neural bypass function might be gradually reduced or removed to allow the neural allograft to take full control of the required functions.

3.2 Method and system for restoring communication in the central nervous system

Let us start by considering a normal physiological system with afferent, relay, and target regions. These could be, for example, the trigeminal nucleus (or ganglia), the sensory thalamus, and the primary somatosensory cortex. As presented in figure 3.1(A), the afferent region sends information to the relay region for processing, after which it is

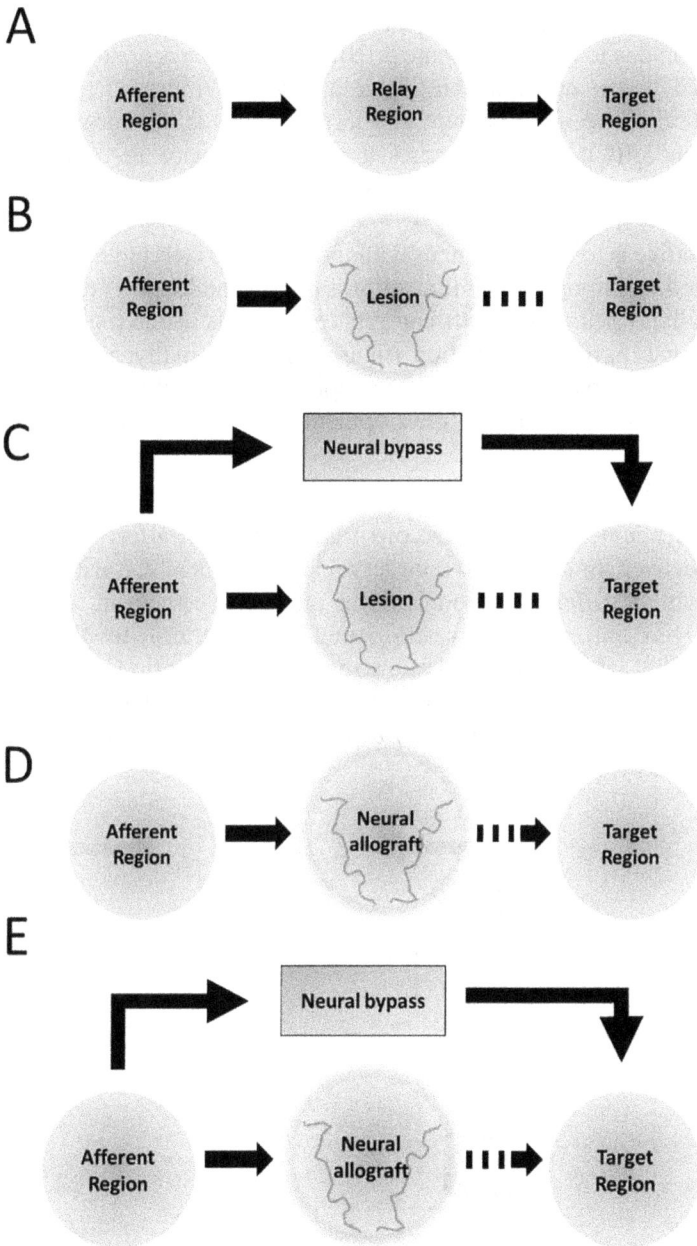

Figure 3.1. System consisting of a combined neural bypass and neural allograft. (A) In the normal nervous system, three regions are sequentially arranged, and information is transferred from the afferent to the relay and then to the target regions. (B) Once a lesion forms along the pathway to/from the relay region, information is no longer transferred to the target region. (C) A neural bypass records the neural activities from the afferent regions, bypasses the lesion, and transfers information to the target region. (D) Hypothetically, a neural allograft can replace the cells in the lesion with new cells and thus restore a particular function. (E) Example of a system that combines a neural bypass and an allograft.

transmitted to the target region. Once damage has occurred along the pathway to/from the relay region due to a lesion/s (figure 3.1(B)), information is no longer processed and/ or transferred to the target region. Thus, a potential approach to restore effective communication between the afferent and target regions is the integration of a neural bypass (figure 3.1(C)). This comprises a prosthetic device that can record neural activities from the afferent region, bypass the lesion, and deliver the activities to the target region (Guggenmos *et al* 2013).

An alternative is to employ a neural allograft, where cells (e.g. mesenchymal stem cells) are used to improve communication through the lesion (Teixeira *et al* 2013). Hypothetically, a neural allograft can replace the cells in the lesion with new cells, thus restoring a particular function (figure 3.1(D)). Finally, in figure 3.1(E), the combined neural bypass and allograft is depicted. The goal is to temporarily utilize the neural bypass to support the development of the neural allograft.

3.3 Neural bypass

In phase I, a neural bypass will record signals from an afferent region, bypass a lesion, and deliver stimulation to a target region. Specifically, it will record neuronal activities from the afferent region, analyze and decode them in real time (using a decoding algorithm), and deliver a stimulation pattern to the target region (figure 3.2). This arrangement will allow for the circumvention of the lesioned relay region, as previously described (Guggenmos *et al* 2013). *This initial phase of the system will have a single outcome that depends on sequential information processing from the neural bypass.*

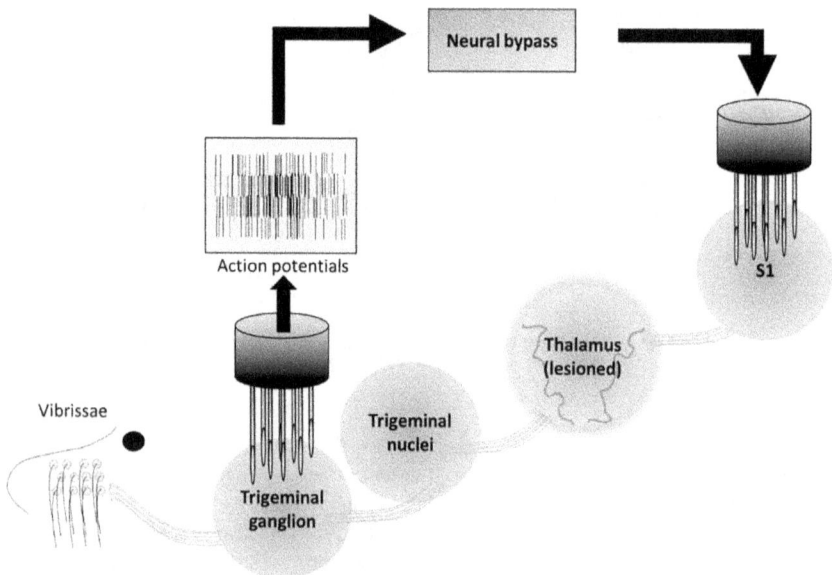

Figure 3.2. Neural bypass (using the example of a rodent nervous system): neural activity recorded from the afferent region (trigeminal ganglion) is decoded in real time and sent as stimulation to a target region (S1), thus bypassing the lesion in the thalamus.

The correct functioning of the neural bypass can be confirmed by the ability of the subject to perform a function or exhibit a particular behavior when the system comprising the neural bypass is engaged, but not when it is disengaged. In addition, recording and analyzing the signals from the target region should enable the identification of the structure initially stimulated (e.g. the whisker of a test rodent).

3.4 Scaffold formed from microcolumns

In phase II, a scaffold that is spatially divided into microcolumns will support neuronal grafts that can be stimulated and recorded (figure 3.3(A, B)). These microcolumns will be spatially organized to reflect the structural properties and spatial organization of the original relay region (figure 3.4).

This arrangement is important because neuronal computational properties should remain distinct within each different microcolumn. The extent to which this spatial organization is determined may vary, depending on the role of the original region. For example, if the goal is to maintain very clear spatial boundaries (e.g. identification of individual rodent whiskers that are stimulated), then it may be desirable to maintain independent microcolumns. Otherwise, if a lower degree of spatial isolation is desired, for example, the microcolumns may support variable levels of communication with adjacent structures (porous microtubules) to promote synaptic arrangements with neighboring columns or proximal regions (i.e. sensory integration). *This phase of the system will have a single outcome that depends on spatial and sequential information processing performed by the neurons present in the scaffold microcolumns.*

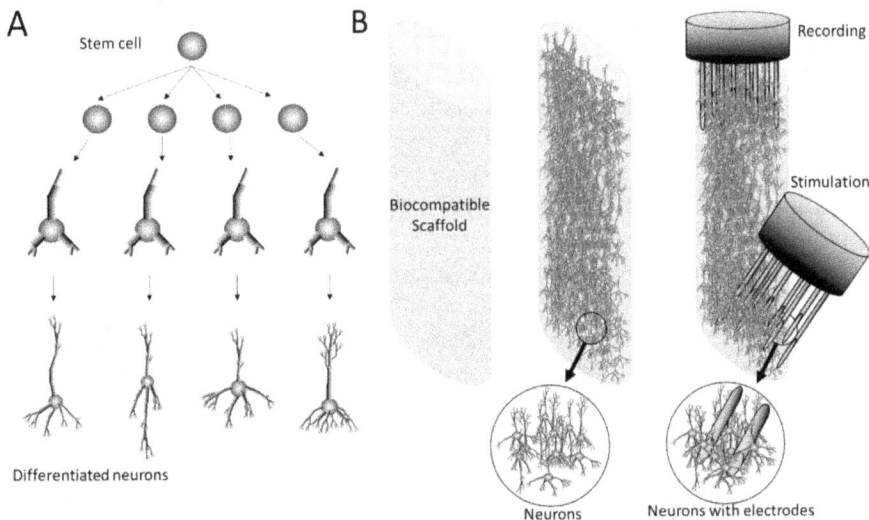

Figure 3.3. Neural scaffold. (A) A graft of neurons derived from mesenchymal cells will be placed in a scaffold (B) to create a microcolumn. Neural scaffolds may be composed of multiple microcolumns (not shown, see figure 3.4) to reflect the spatial organization of specific original neural regions and pathways. Scaffolds can also include stimulating and recording electrodes to allow the modulation of graft neurons.

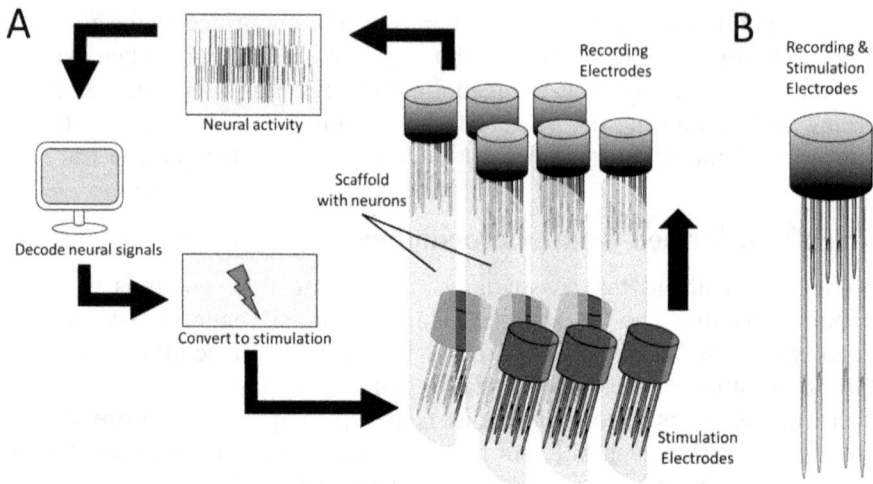

Figure 3.4. Neural graft testing. (A) Graft of spatially arranged microcolumns (blue cylinders) with stimulation patterns that reflect the neural activity recorded from the subject being delivered to neurons located at the bottom of the scaffold. Recordings are made from neurons located at the top of the scaffold. (B) Alternative presentation of the arrangement of recording and stimulating electrodes.

The correct functioning of the neural scaffold should be revealed by its ability to stimulate and record the neural activities of neurons in the microcolumns (figure 3.4). Furthermore, the stimulation of different microcolumns (i.e. different spatial locations) should be correlated with the capacity to decode different spatial patterns of neural activities from the recordings.

3.5 Remote training of the neural graft

In phase III, the signals recorded from the afferent region will be duplicated, with the first copy following the neural bypass (as previously), while the second copy will be sent to the neural graft. The scaffold supporting the neural allografts (placed in a remote location) will receive stimulation patterns that reflect the same spatial organization as that found in the original lesioned region (figure 3.5). This stimulation will be delivered at one end of the microcolumns (e.g. the bottom neurons), and neural responses will be recorded from the other end (e.g. the top neurons). For example, if the afferent signals are known to originate from tactile stimuli applied to three different whiskers of a rodent, each microcolumn would receive stimulation associated with these distinct whiskers. The recording and analysis of the signals from the other end of the microcolumns (e.g. the top neurons) should allow for the identification of the structures that were initially stimulated (i.e. individual whiskers). It is noteworthy that for this phase only, the outcomes of the neural bypass will be used for the functional/behavioral purposes of the subject. *For this phase, the complete system will have two distinct parallel outcomes: one relating to the neural bypass and another relating to the neural allografts.*

The correct functioning of the neural bypass should be revealed by the ability of the subject to perform a function or exhibit a behavior when the neural bypass is

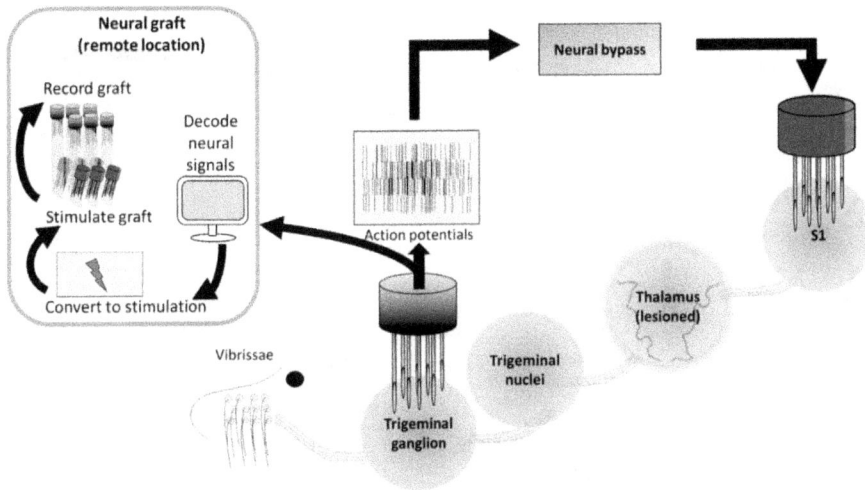

Figure 3.5. Neural graft training with real-time data. A graft of spatially arranged microcolumns (box) located in a remote location will receive stimulation patterns that reflect the neural activity recorded from the subject using the neural bypass (note the bifurcation of the signal following neural activity recorded in the trigeminal ganglion). Data from the neural bypass and the neural graft are processed independently. The results are then compared to identify the learning features in the neural allograft.

engaged, but not when it is disengaged. Additional recordings of the neural activities in the target area should support the behavioral findings. Furthermore, the appropriate functioning of the scaffold should be revealed by the ability of the system to decode the spatial organization of the delivered stimuli based solely on the outputs of the scaffold neurons (e.g. the top neurons).

3.6 Remote control of the neural bypass by a remote neural graft

In phase IV, signals acquired from the afferent region of the neural bypass will be diverted and processed by the trained scaffold in a remote location (figure 3.6) and then sent again to the target region in the nervous system of the subject. Thus, neural activity recorded from the afferent region of the subject will be processed in the microcolumns of the neural scaffold, with the outcomes being used to stimulate neurons in the target region. This arrangement forces the flow of information to go through an additional stage where the trained microcolumns determine which stimulation patterns are delivered to the target region. *In this phase, the complete system will have one single outcome that depends on sequential information processing from the neural bypass to the neural scaffold and then again to the neural bypass.*

The correct functioning of the neural bypass should be revealed by the ability of the subject to perform a function or exhibit a behavior when the neural bypass and neural scaffold are communicating, but not when either or both are not. Additional recordings from neural activities in the neural scaffold and in the target area should also support the behavioral findings (i.e. enable the identification of stimuli applied to the subject). Furthermore, the correct functioning of the scaffold should be

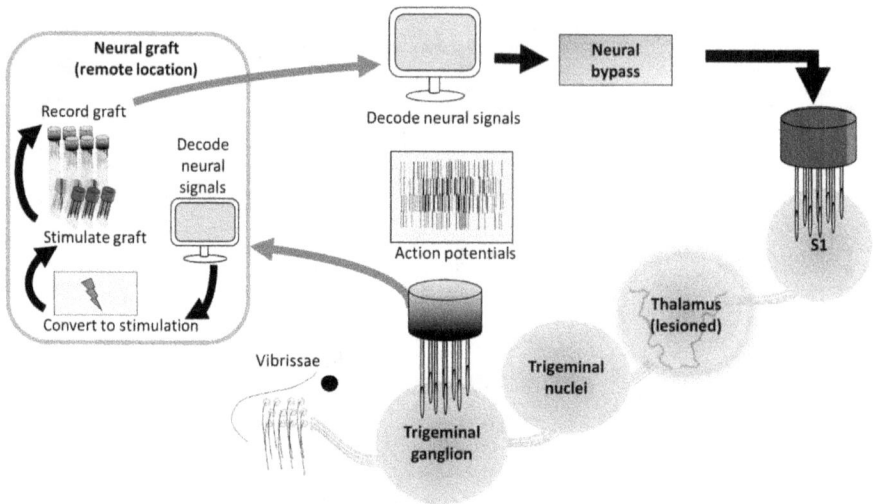

Figure 3.6. Neural allograft processing information in a remote location. Signals recorded from the afferent region (trigeminal ganglion) will be sent to a remote location where the neural allograft will process the relevant information. Stimulation (red electrodes) of the bottom neurons of the microcolumn will modulate the graft neurons. After recording the graft neural activity (black electrodes on top of the microcolumns), information will then be sent to the target area (S1). In this phase, the outcomes of the neural activities performed by the graft will determine the stimulation patterns delivered at the end of the neural bypass.

revealed by the ability of the system to decode the spatial organization of the delivered stimuli based solely on the outputs of scaffold neurons (e.g. top neurons).

3.7 Implantation of the trained neural graft

In phase V, the scaffold of neurons previously trained and tested is now implanted into the nervous system of the subject. In this phase, no changes are made in the way information is processed. The only difference relative to phase IV is that the scaffold of neurons is now subject to the chemical, physical, and computational constraints of the surrounding tissue. In addition, this means that, depending on the characteristics of the scaffold, there is the potential for these neurons to establish additional connections with surrounding structures and develop additional computational properties. *In this phase, the complete system will have one single outcome that depends on information processing from both the neural bypass (figure 3.7) and the neural scaffold, but also from the interactions between the neural scaffold and surrounding structures.*

The correct functioning of the neural bypass should be revealed by the ability of the subject to perform a function or exhibit a behavior when the neural bypass and neural scaffold are communicating (i.e. the system is engaged), but not when either or both of them are not (i.e. the system is disengaged). Additional recordings from neural activities in the neural scaffold and in the target area should support the behavioral findings (i.e. enable the identification of stimuli applied to the subject).

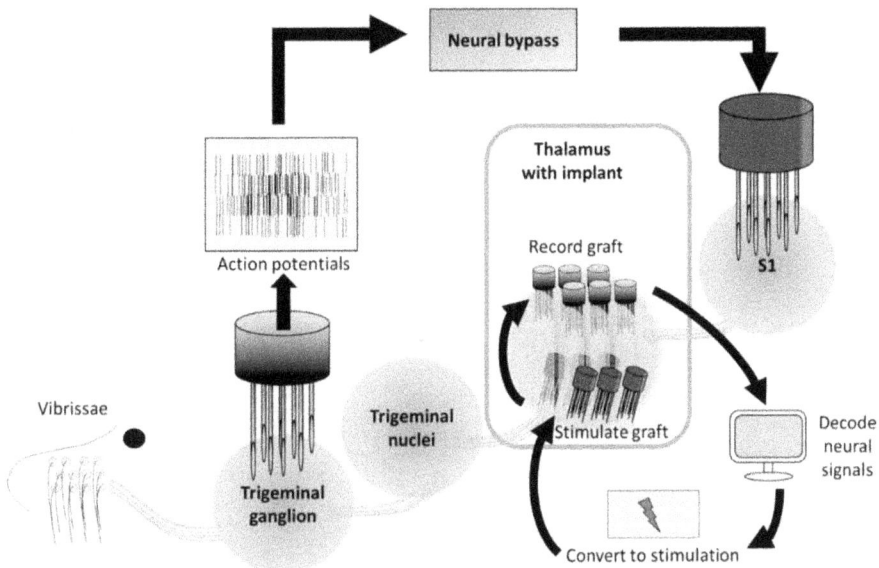

Figure 3.7. Neural bypass and neural grafts implanted in the subject. After testing and training, the neural allograft is implanted in the location of the lesioned region (here represented as the thalamus). Information processing remains the same as in the previous phase, but the computations and survival of the neural graft are now also influenced by localized environmental factors.

Furthermore, the correct functioning of the scaffold should be revealed by the ability of the system to decode the spatial organization of the delivered stimuli based solely on the outputs of scaffold neurons (e.g. top neurons). Finally, the quality and number of neurons recorded, as well as the changes in neural population activities, should allow the survival status of the neural graft to be monitored.

3.8 Gradual removal of the neural bypass

In phase VI, the neural bypass will be gradually disconnected to promote a lead role for the neural allograft. The only difference between this phase and the preceding one is that the responsibility for function now belongs to the neuronal scaffold. The goal is to promote the complete restoration of neural functionality through the gradual removal of the neural bypass (figure 3.8). It is unlikely that the neural graft will lead to neural connections that are similar to those present in the original region it intends to replace. However, due to the continuous reward effect of the combined neural bypass and neural scaffold, it is expected that alternative neural pathways may be used by the neuronal scaffold. Thus, the gradual disconnection of the neural bypass to increase reliance on the neuronal allograft may allow for the partial or complete removal of the functions of the neural bypass. *In this phase, the complete system will have a single outcome that depends more on the information processing from the neural allograft and less on the neural bypass.*

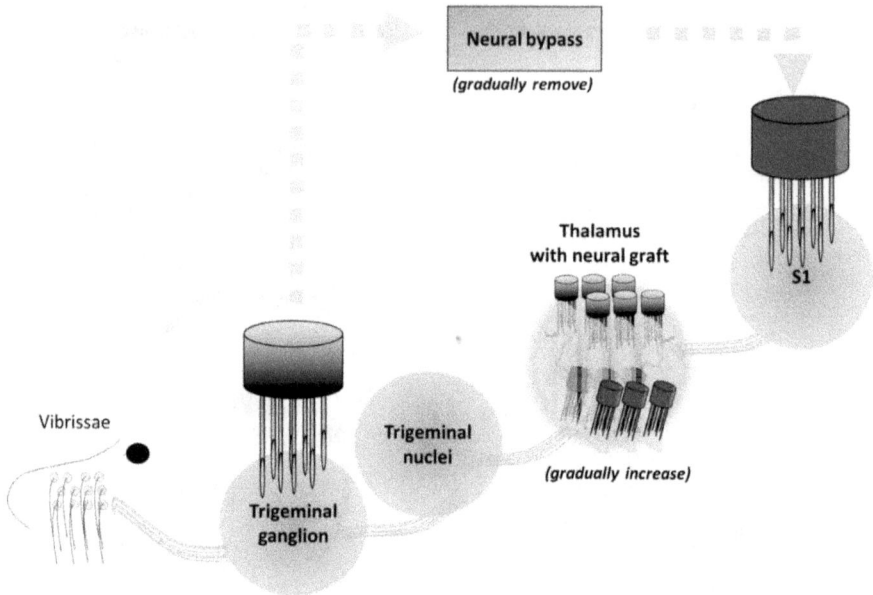

Figure 3.8. Neural bypass withdrawal. Depending on the region and the evolution of the individual subject, the neural bypass may be gradually disconnected to promote reliance on the neural graft. This may change the computations performed in the neural graft or facilitate the use of alternative neural pathways.

The correct functioning of the neural bypass should be revealed by the ability of the subject to perform a function or exhibit a behavior when the neural bypass has been disconnected (i.e. the neural bypass is disengaged) and the neural allograft is allowed to operate freely. Additional recordings from the neural activities of the neuronal scaffold and target area should also support the behavioral findings (i.e. allow for the identification of stimuli applied to the subject). Furthermore, the correct functioning of the scaffold should be revealed by the ability of the system to decode the spatial organization of the delivered stimuli based solely on the outputs of the scaffold neurons (e.g. top neurons).

Finally, the quality, number of neurons recorded, and changes in neural populations should allow for the survival of the neural graft to be evaluated.

3.9 Conclusions

We presented a potential approach for the development of a tactile neuro-prosthetic thalamus. This method involves a series of sequential steps where two different techniques are combined. The current strategy is aimed at a proof of concept. Consequently, the experiments presented here were designed primarily for rodents, where the anatomical organization of the somatosensory system (Watson 2012, Petersen 2019) is likely to facilitate the physical separation of information.

Definitions

Afferent region: the first region considered in this system (for example, sensory afferent).

Target region: region that will receive information processed by the system (neural bypass and/or neural allograft).

Relay region: region that receives information from the afferent region, processes it, and transfers it to the target region (may also compute additional functions).

Recording or stimulating electrode: physical devices used to record information related to neural activity and modulate neural processes (may include electrical stimulation, transcranial magnetic stimulation, and others).

Neural graft: neuron culture implanted in the nervous system of the subject (may be derived from stem cells or others).

Microtubules: biocompatible structures that have the potential to support cultured neurons.

Microcolumn: a microtubule with cultured neurons within it.

Scaffold: structure that comprises one or more microtubules (defines the spatial arrangement of microtubules).

Neural bypass: a neural prosthetic device that bridges lesions within the nervous system.

Stimulator: device that generates electrical current or other effects (e.g. magnetic fields) to modulate neural activities.

Remote location: located outside the nervous system.

References and further reading

Alishahi M, Anbiyaiee A, Farzaneh M and Khoshnam S E 2020 Human mesenchymal stem cells for spinal cord injury *Curr. Stem Cell Res. Ther.* **15** 340–8

Bouton C E 2019 Restoring movement in paralysis with a bioelectronic neural bypass approach: current state and future directions *Cold Spring Harb. Perspect. Med.* **9** a034306

Cicchetti F, Saporta S, Hauser R A, Parent M, Saint-Pierre M, Sanberg P R *et al* 2009 Neural transplants in patients with Huntington's disease undergo disease-like neuronal degeneration *Proc. Natl. Acad. Sci. U. S. A.* **106** 12483–8

Cisbani G, Maxan A, Kordower J H, Planel E, Freeman T B and Cicchetti F 2017 Presence of tau pathology within foetal neural allografts in patients with Huntington's and Parkinson's disease *Brain* **140** 2982–92

de Araújo L T, Macêdo C T, Damasceno P K F, das Neves Í G C, de Lima C S, Santos G C *et al* 2022 Clinical trials using mesenchymal stem cells for spinal cord injury: challenges in generating evidence *Cells* **11** 1019

Dlouhy B J, Awe O, Rao R C, Kirby P A and Hitchon P W 2014 Autograft-derived spinal cord mass following olfactory mucosal cell transplantation in a spinal cord injury patient: case report *J. Neurosurg, Spine.* **21** 618–22

Donati A R, Shokur S, Morya E, Campos D S, Moioli R C, Gitti C M *et al* 2016 Long-term training with a brain–machine interface-based gait protocol induces partial neurological recovery in paraplegic patients *Sci. Rep.* **6** 30383

Guggenmos D J, Azin M, Barbay S, Mahnken J D, Dunham C, Mohseni P and Nudo R J 2013 Restoration of function after brain damage using a neural prosthesis *Proc. Natl. Acad. Sci. U. S. A.* **110** 21177–82

Hebb D O 2002 *The Organization of Behavior: A Neuropsychological Theory* (New York, NY: Psychology Press)

Horner P J and Gage F H 2000 Regenerating the damaged central nervous system *Nature* **407** 963–70

Illis L S 2012 Central nervous system regeneration does not occur *Spinal Cord* **50** 259–63

Jensen G, Holloway J L and Stabenfeldt S E 2020 Hyaluronic acid biomaterials for central nervous system regenerative medicine *Cells* **9** 2113

Lebedev M A and Nicolelis M A 2017 Brain–machine interfaces: from basic science to neuro-prostheses and neurorehabilitation *Physiol. Rev.* **97** 767–837

Lima C, Pratas-Vital J, Escada P, Hasse-Ferreira A, Capucho C and Peduzzi J D 2006 Olfactory mucosa autografts in human spinal cord injury: a pilot clinical study *J. Spinal Cord Med.* **29** 191–203 discussion 204-6

Marsh S E and Blurton-Jones M 2017 Neural stem cell therapy for neurodegenerative disorders: the role of neurotrophic support *Neurochem. Int.* **106** 94–100

Pais-Vieira M, Lebedev M, Kunicki C, Wang J and Nicolelis M A 2013 A brain-to-brain interface for real-time sharing of sensorimotor information *Sci. Rep.* **3** 1319

Pais-Vieira M, Chiuffa G, Lebedev M, Yadav A and Nicolelis M A 2015 Building an organic computing device with multiple interconnected brains *Sci. Rep.* **5** 11869

Pais-Vieira C, Gaspar P, Matos D, Alves L P, da Cruz B M, Azevedo M J *et al* 2022 Embodiment comfort levels during motor imagery training combined with immersive virtual reality in a spinal cord injury patient *Front. Hum. Neurosci.* **16** 909112

Petersen C C H 2019 Sensorimotor processing in the rodent barrel cortex *Nat. Rev. Neurosci.* **20** 533–46

Rowald A, Komi S, Demesmaeker R, Baaklini E, Hernandez-Charpak S D, Paoles E *et al* 2022 Activity-dependent spinal cord neuromodulation rapidly restores trunk and leg motor functions after complete paralysis *Nat. Med.* **28** 260–71

Selfslagh A, Shokur S, Campos D S F, Donati A R C, Almeida S, Yamauti S Y *et al* 2019 Non-invasive, brain-controlled functional electrical stimulation for locomotion rehabilitation in individuals with paraplegia *Sci. Rep.* **9** 6782

Sharma G, Friedenberg D A, Annetta N, Glenn B, Bockbrader M, Majstorovic C *et al* 2016 Using an artificial neural bypass to restore cortical control of rhythmic movements in a human with quadriplegia *Sci. Rep.* **6** 33807

Shokur S, Donati A R C, Campos D S F, Gitti C, Bao G, Fischer D *et al* 2018 Training with brain–machine interfaces, visuo-tactile feedback and assisted locomotion improves sensor-imotor, visceral, and psychological signs in chronic paraplegic patients *PLoS One* **13** e0206464

Silva N A, Salgado A J, Sousa R A, Oliveira J T, Pedro A J, Leite-Almeida H *et al* 2010 Development and characterization of a novel hybrid tissue engineering-based scaffold for spinal cord injury repair *Tissue Eng. Part* A **16** 45–54

Teixeira F G, Carvalho M M, Sousa N and Salgado A J 2013 Mesenchymal stem cells secretome: a new paradigm for central nervous system regeneration *Cell. Mol. Life Sci.* **70** 3871–82

Tsintou M, Dalamagkas K and Seifalian A M 2015 Advances in regenerative therapies for spinal cord injury: a biomaterials approach *Neural Regen. Res.* **10** 726–42

Varadarajan S G, Hunyara J L, Hamilton N R, Kolodkin A L and Huberman A D 2022 Central nervous system regeneration *Cell* **185** 77–94

Watson C 2012 The somatosensory system *The Mouse Nervous System* 563–70 (New York: Academic)

Williams A 2014 Central nervous system regeneration—where are we *QJM* **107** 335–9

Xue W, Shi W, Kong Y, Kuss M and Duan B 2021 Anisotropic scaffolds for peripheral nerve and spinal cord regeneration *Bioact. Mater.* **6** 4141–60

Zhang Q, Shi B, Ding J, Yan L, Thawani J P, Fu C and Chen X 2019 Polymer scaffolds facilitate spinal cord injury repair *Acta Biomater.* **88** 57–77

Chapter 4

Neurotech frontiers: ethics, identity, and the evolution of brain–computer interfaces

Ingrid Vasiliu-Feltes

This chapter provides an overview of neurotechnology's ethical intricacies and a deeper examination of brain–computer interfaces (BCIs), which may be extrapolated in many respects to hypothetical (for now) brain/cloud interfaces (B/CIs) (Martins *et al* 2019), emphasizing the need for comprehensive ethical frameworks that guide their design, deployment, and societal integration. As BCIs advance toward becoming integral to human augmentation and neurological interventions, this chapter explores critical ethical dimensions, extending from digital ethics to governance, with insights into deep tech convergence and strategic leadership roles in ensuring their responsible development.

> *Section I: Ethical Foundations in Neurotechnology and BCIs* lays the groundwork by addressing various ethical frameworks applicable to BCIs, starting with digital ethics, which covers the evolving responsibilities and considerations in a connected, data-centric landscape. Bioethics addresses the moral implications of neurointerventions on human biology, whereas medical ethics examines patient autonomy, safety, and the potential for misuse in medical settings. This chapter also touches on societal ethics, exploring how neurotechnology might reshape societal structures, and corporate ethics, evaluating the responsibilities of corporate actors in ensuring that neurotechnology aligns with the social good. Finally, the section on applied ethics surveys current research trends, emphasizing practical ethical applications in ongoing neurotechnology projects.

> *Section II: Redefining Digital Identity in BCIs* introduces the complexities of digital identity as BCIs blur the lines between biological and digital existence. Given the evolving nature of BCI functionalities, the discussion on dynamic informed consent emphasizes the need for flexible, ongoing consent frameworks in neurotechnology. Furthermore, this section explores rights and citizenship in a future where neuroaugmented individuals may seek new legal

4-1

and digital citizenship forms. Governance models are proposed to ensure ethical oversight of these emerging digital identities, protecting individual rights and societal harmony.

Section III: Ethical Considerations in Hybrid Augmented BCIs highlights the transformative impact of BCIs on learning, education, and employment. Hybrid BCIs, which merge cognitive augmentation with external systems, raise ethical questions about equitable access, cognitive enhancement disparities, and the transformation of traditional learning and employment paradigms. This section emphasizes the need for ethical guidelines to ensure these technologies benefit all societal strata without reinforcing existing inequalities.

Section IV: BCI Harmonization addresses the integration of BCIs with existing socio-technical systems, emphasizing cyberethics as a crucial foundation for safeguarding privacy, security, and digital rights. Sustainable BCIs explore the ethical considerations of creating long-term, eco-conscious BCI systems. In contrast, inclusive and diverse BCIs address the need for neurotechnological solutions that reflect the diversity of users, avoiding cultural and demographic biases in design and implementation.

Section V: Deep Tech Convergence examines the intersections between BCIs and emerging technologies (e.g. artificial intelligence (AI), digital twins, blockchain, bioimplants, multicloud computing, 6G, and satellite internet). This section explores how these converging technologies can complement BCIs and introduce additional ethical complexities, particularly as they relate to data ownership, interoperability, and global digital equity.

Section VI: Strategic Considerations concludes by outlining the ethical roles of key leadership tiers (e.g. boards, C-suite executives, and management) in overseeing the ethical implementation and governance of BCIs. By embedding ethical considerations into their strategic decision-making processes, these leaders can guide their organizations to foster innovation while ensuring ethical integrity and societal benefit.

This chapter is an essential read for engineers, technologists, ethicists, and corporate leaders involved in neurotechnology, as it offers a comprehensive and strategic view of ethical considerations for BCIs.

4.1 Introduction

This chapter presents a comprehensive and systematic exploration of the critical ethical frameworks that govern the development and deployment of neurotechnology, with a focused examination of BCIs. A robust ethical foundation becomes paramount as neurotechnology advances toward dynamic, complex, and intimate interactions with human cognition and neural systems. To this end, this chapter commences with an analysis of foundational ethical principles, including respect for autonomy, beneficence, non-maleficence, justice, equity, and solidarity, which form the backbone of ethical deliberation in neurotechnology.

These ethical constructs are not merely theoretical. Instead, they provide the basis for guiding the practical application and governance of BCIs to ensure their responsible integration into human cognitive, medical, and societal domains. The ethical imperatives of respect for persons and autonomy emphasize the need for safeguarding individual rights and freedoms in the context of neurotechnological enhancement. Meanwhile, beneficence and non-maleficence compel stakeholders to optimize positive outcomes and minimize harm in implementing BCIs, particularly given their potential for deep integration with cognitive processes. Furthermore, justice, equity, and solidarity underscore the importance of equitable access to neurotechnological advancements, ensuring that these technologies are inclusive and accessible to all rather than reinforcing societal inequalities.

The chapter progresses into a specialized discussion on the convergence of BCIs with other emerging technologies, which emphasizes the need for a customizable, adaptive approach to neurotechnology. In this context, the discussion introduces two proprietary terms, Precision BCIs™ and BCI-Powered Hybrid Augmented Intelligence™, which represent the forefront of personalized neurotechnological systems. Precision BCIs™ refer to neural interface solutions that are meticulously calibrated to align with each user's individual cognitive and neurological profiles. These personalized systems are not designed for mass application; instead, they represent a paradigm shift towards individualized, adaptive neurotechnology solutions that dynamically respond to each user's unique cognitive, emotional, and physical characteristics.

In alignment with these principles, Precision BCIs™ extend the philosophy of precision medicine into the domains of cognitive and neurological augmentation, ensuring that neurotechnological interfaces are tailored to the individual's neuro-biological framework, learning modalities, and executive functioning patterns. By embedding neuroadaptive algorithms, these BCIs continuously adjust their functionality to optimize user interactions and enhance cognitive performance, maximizing usability and the ethical application of such interfaces.

BCI-Powered Hybrid Augmented Intelligence™ further extends the possibilities of neurotechnology by facilitating the seamless integration of BCIs into advanced AI systems. This synergy between human cognition and AI leads to the creation of hybrid systems that augment human capabilities across multiple domains, including cognitive processing, decision-making, and adaptive learning. The hybrid augmented intelligence framework fosters a more dynamic interaction between the user and neurotechnology, where real-time cognitive feedback loops enable the technology to learn and adapt to the user's needs, environment, and context of use. This continuous, bidirectional exchange of data and feedback underpins the system's ethical architecture to ensure that it remains responsive and transparent to the user.

This chapter advocates for these personalized and hybrid solutions and foregrounds the ethical responsibilities inherent in their design and implementation. Integrating BCIs with AI demands the re-evaluation of consent, privacy, and data governance, as these systems generate immense quantities of sensitive neurological data. Therefore, the ethical architectures of BCI systems must include stringent protocols for dynamic, informed consent, ensuring users maintain autonomy over their data and neural augmentation processes throughout the technology's lifecycle.

The ethical considerations surrounding these technologies are not confined to their technical implementation. The core focus here involves how BCIs impact organizational workflows, especially in hybrid augmented intelligence environments. This includes an exploration of the strategic and ethical challenges faced by different organizational tiers (e.g. boards, C-suite executives, management, and operational teams) that are tasked with deploying and overseeing neurotechnological systems. By dissecting the roles and responsibilities at each level, a roadmap is provided for embedding complex ethical frameworks into decision-making processes, from high-level governance to day-to-day operational management.

For boards and C-suite executives, ethical governance of neurotechnology involves aligning corporate objectives with societal good and ensuring the long-term sustainability of neurotechnology applications. These leaders must navigate the evolving regulatory landscape, anticipate public concerns around privacy and data security, and ensure that their organizations uphold the highest standards of corporate ethics in neurotech innovation. A focus on strategic foresight and ethical risk mitigation is paramount, as boards and executives are responsible for guiding their organizations through the ethical complexities of advanced neurotechnologies.

For management and operational teams, ethical considerations are more practical and tactical. The implementation of BCI-powered hybrid augmented workflows presents unique challenges, such as maintaining equity of access to augmentation technologies, preventing bias in AI algorithms integrated with BCIs, and ensuring the ethical use of neurotechnological data in decision-making processes. Managers are tasked with translating high-level ethical principles into operational guidelines, fostering ethical innovation, and ensuring that neurotechnology advances equitably benefit a broad spectrum of stakeholders.

By offering a detailed analysis and practical guidance, this chapter aims to provide stakeholders with the tools required to navigate the ethical landscape of neurotechnology, particularly in the realm of BCIs. The discussion of hybrid augmented workflows highlights the opportunities presented by these technologies and emphasizes the ethical imperatives that must be adhered to in their deployment. The comprehensive approach taken in this chapter fosters a deeper understanding of how ethical frameworks can be embedded in the development and deployment of neurotechnology, ensuring that technological advancements are aligned with societal values and contribute positively to the broader social fabric.

The ethical challenges of BCIs and hybrid augmented intelligence systems are systematically addressed below, offering a forward-thinking, interdisciplinary perspective on how these technologies may be developed and deployed in ways that respect individual autonomy, promote equity, and foster inclusive societal progress. The framework provided here is essential for ensuring that neurotechnology serves the greater good while upholding the highest standards of ethical integrity.

4.2 Section I: ethical foundations of neurotechnology and brain–computer interfaces

4.2.1 Digital ethics in neurotechnology: an advanced examination

The advent of neurotechnology presents profound ethical challenges that must be addressed through a robust and dynamic framework of digital ethics. As BCIs and other neurotechnological systems evolve, their interactions with cognitive and sensory functions will call for an intricate set of ethical guidelines to ensure safety and fairness. The cornerstone of such a framework must be grounded in the ethical principles that safeguard human dignity, privacy, and individual autonomy while promoting equitable access, spanning all social and economic strata, to these groundbreaking technologies.

Respect for persons is a foundational tenet in digital ethics that demands a thorough examination of user autonomy and privacy. In neurotechnology, this principle extends far beyond traditional notions of consent. Neurotechnologies collect and process highly sensitive neurodata—data derived from brain activity, which is deeply personal and rich in cognitive insights. Users must retain control over this data, its interpretation, and its applications. The concept of dynamic, informed consent is central to this principle, as it ensures that consent is not static but evolves with neurotechnology's changing functionalities and capabilities. This continuous and iterative consent process must be embedded within system design, enabling users to provide or revoke consent at any point, particularly as new applications and data uses emerge.

The principle of respect for persons also extends to ensuring transparency in how neurodata is processed and utilized by various stakeholders, including healthcare providers, corporations, and researchers. Clear and unambiguous communication is essential in articulating the data that is being collected, how it is stored, who has access to it, and how it is intended to be used. Transparency must be operationalized through user-friendly interfaces and robust data governance policies, ensuring that individuals fully understand the scope and implications of their participation in neurotechnological systems.

Beneficence underscores the ethical obligation to ensure that neurotechnologies are designed and deployed to enhance the cognitive and sensory capacities of users. In practical terms, this principle mandates that all neurotechnological interventions and applications must provide measurable benefits to the user, whether in the form of cognitive enhancement, restoration of sensory functions, or therapeutic applications. Adhering to this principle requires a commitment to rigorous testing and validation procedures during the research and development phases to ensure that the risks are minimal and significantly outweighed by the benefits. The principle of beneficence also extends to continuous postmarket surveillance of neurotechnological devices. Given potentially unintended consequences as users interact with these systems over time, developers and regulators must establish feedback mechanisms that monitor real-world usage and address emerging issues. These mechanisms should include regular updates to the algorithms and software that govern neurotechnological systems, ensuring that they evolve in response to user experiences and advancements in scientific understanding.

Non-maleficence is a critical principle that obliges developers to prevent harm to users by all means. This principle takes on heightened importance in neurotechnology, given the direct interaction of these systems with the human brain. Several layers of protection must be incorporated into the ethical design of BCIs and other neurotechnologies. First, personal neurodata must be safeguarded from unauthorized access or malicious use. The growing sophistication of cyber threats necessitates the development of advanced cybersecurity protocols (e.g. nested randomly refreshing quantum encryption) that protect neurodata from breaches, theft, and manipulation. End-to-end encryption, multifactor authentication, and decentralized data storage architectures are essential for a secure neurotechnology ecosystem.

Additionally, the algorithms that power neurotechnologies must undergo rigorous testing and validation before being deployed in real-world environments. These additional measures include testing for accuracy and efficacy and ensuring that the algorithms are free from bias, do not generate harmful or misleading outputs, and maintain reliability under diverse usage conditions. To maintain trust and integrity, these testing protocols should adhere to the highest standards set by bodies such as IEEE and NIST, ensuring consistency in evaluation criteria and methodologies.

Justice within digital ethics calls for equitable access to neurotechnology across diverse populations. This principle highlights the responsibility of developers, policymakers, and stakeholders to ensure that the benefits of neurotechnology are distributed fairly and without discrimination. Socioeconomic disparities should not preclude individuals from accessing neurotechnological innovations, and proactive measures must be taken to eliminate barriers that limit access for underrepresented groups.

The principle of justice is particularly salient in neurotechnology applications in healthcare, where there is a risk that these advanced systems could exacerbate existing inequalities. To address this risk, ethical frameworks should include provisions for affordability and availability, ensuring that neurotechnologies are integrated into public health systems and are accessible to marginalized communities. Regulatory bodies and industry leaders must collaborate to develop policies that promote equal access and address any financial or systemic hurdles that may prevent certain groups from benefiting from neurotechnological advancements.

Closely related to justice is the principle of equity, which expands the notion of fairness to include efforts to actively dismantle systemic barriers that prevent individuals from accessing and benefiting from neurotechnology. Equity-driven frameworks seek to address disparities in access and differences in how individuals experience neurotechnology. Diverse populations may have different neurological, cultural, or social needs, and neurotechnological systems must be designed to accommodate these variations. An inclusive approach to research and development, where diverse user groups are involved in the design and testing phases, can ensure that neurotechnology is responsive to unique requirements.

Furthermore, equity demands that developers continuously evaluate and adjust their systems to prevent any group from being disproportionately burdened by negative outcomes or unintended consequences. Ethical frameworks should promote adaptive algorithms that can learn from diverse datasets and avoid the replication of biases that are present in traditional technologies. This adaptability will ensure that neurotechnological systems remain inclusive and equally benefit users from all backgrounds.

Solidarity represents the collective ethical responsibility of all stakeholders (e.g. developers, users, researchers, regulators, and corporations) to work together to advance neurotechnologies in ways that prioritize the common good while safeguarding individual rights. In digital ethics, solidarity promotes a cooperative approach to data sharing and collaboration that benefits broader society rather than individual or corporate interests alone.

For instance, the sharing of neurodata between research institutions, healthcare providers, and developers can significantly advance the understanding and treatment of neurological conditions. However, such data sharing must be governed by strict ethical guidelines that protect privacy and ensure that individuals' rights are not compromised. Federated learning techniques, where data is shared without compromising individual privacy, offer a promising solution to this challenge, allowing for large-scale collaboration without sacrificing individual autonomy.

Solidarity also extends to regulatory and governance structures. Ethical frameworks must encourage collaboration between national and international regulatory bodies to establish consistent neurotechnology standards. These standards should reflect a shared commitment to advancing neurotechnology in ways that are safe, effective, and accessible to all while also respecting the unique legal and cultural contexts of different regions.

4.2.2 Societal ethics in neurotechnology: a holistic approach

The rapid advancement of neurotechnologies, particularly BCIs, necessitates a comprehensive ethical framework that addresses the societal implications of these innovations. Societal ethics extends beyond individual considerations, as it focuses on the broader impacts of neurotechnologies on communities, public welfare, and the global social fabric. By emphasizing principles such as respect for persons, beneficence, non-maleficence, justice, equity, and solidarity, societal ethics in neurotechnology promotes responsible development that ensures inclusivity, fairness, and collective well-being.

At the core of societal ethics resides the principle of respect for persons, which underscores the importance of safeguarding privacy and autonomy within society. Neurotechnologies, by their very nature, collect and analyze vast amounts of personal neural data, which raises significant concerns regarding data privacy and ownership. Respect for persons mandates that individuals retain control over their neurodata, with transparent and robust consent mechanisms that extend beyond individuals to society. This principle also ensures that neurotechnological developments respect the rights of individuals to make autonomous decisions regarding the use of their neural data, free from coercion or manipulation. By safeguarding privacy and autonomy, this principle protects not just the individual but also the integrity of social structures that rely on trust and individual freedoms.

Beneficence in societal ethics extends beyond individual welfare to promote the broader societal good. Neurotechnological advancements must aim to improve the quality of life for the population as a whole, addressing critical societal challenges such as mental health, cognitive disabilities, and neurological disorders. The principle of beneficence requires that neurotechnology should not only serve those directly using it but also contribute to societal well-being. In this context, it calls for the development of neurotechnologies that enhance public health, increase cognitive accessibility, and reduce societal barriers to mental and physical well-being. For instance, deploying BCIs to enhance educational tools, cognitive therapies, or rehabilitation services could benefit society by promoting inclusivity and supporting individuals in overcoming neurological impairments.

Non-maleficence (the duty to avoid harm) becomes particularly important when considering the societal ramifications of neurotechnology deployment. While neurotechnologies hold the potential to enhance cognitive capabilities and address neurological conditions, they also risk exacerbating social inequalities, privacy violations, or even creating new societal challenges, such as the commodification of neurodata. Non-maleficence obliges developers, policymakers, and society at large to ensure that these technologies do not create social divides or contribute to systemic harm. This principle advocates for the careful evaluation and mitigation of risks associated with neurotechnology, emphasizing a precautionary approach that prioritizes societal stability and cohesion over rapid technological deployment.

Justice in societal ethics emphasizes equitable access to neurotechnological advancements. The introduction of neurotechnology can potentially create a technological divide between those who can afford it and those who cannot. This disparity risks deepening societal inequalities, as access to cognitive enhancement or therapeutic neurotechnologies may become a privilege of the wealthy. Justice ensures that the benefits of neurotechnology are fairly distributed, advocating for policies and frameworks that provide equal (ultimately free in the future) access to neurotechnologies, regardless of socioeconomic status. This principle supports the creation of public health initiatives and regulatory frameworks, which ensure neurotechnology remains accessible to underserved and marginalized communities, thereby preventing societal divides.

Equity further reinforces the idea of inclusivity in neurotechnological development. While justice ensures fair distribution, equity goes beyond by actively addressing the specific needs of diverse populations. Neurotechnological innovations must be designed and implemented to account for the varied cultural, social, and cognitive contexts within which they will be used. Equity mandates that all societal groups should benefit from these advancements, with particular attention given to marginalized or disadvantaged communities. This principle ensures that neurotechnology serves the entire population, reduces disparities, and promotes inclusivity in design and deployment.

Solidarity in societal ethics promotes a collective approach to addressing the ethical challenges posed by neurotechnology. As these technologies evolve, their societal impacts will require a unified effort to address potential ethical dilemmas and ensure responsible deployment. Solidarity calls for interdisciplinary collaboration, bringing together technologists, ethicists, policymakers, and the public to navigate the ethical complexities of neurotechnologies. It fosters societal cohesion by encouraging mutual support and collective responsibility, ensuring that the development and deployment of neurotechnological innovations align with the common good.

Societal and corporate ethics differ in their scope and focus, especially in the context of neurotechnology. Societal ethics emphasizes collective well-being and focuses on how neurotechnologies such as BCIs impact broader society. It addresses issues such as equitable access, privacy, social justice, and inclusivity to ensure that neurotechnologies benefit all members of society and do not exacerbate existing inequalities. It also prioritizes the protection of public interests and fosters collaboration among stakeholders to promote ethical and responsible development that enhances societal well-being.

Conversely, corporate ethics centers around the responsibilities of companies in the neurotech industry, balancing profit motives with ethical obligations to consumers, employees, and shareholders. Corporate ethics in neurotechnology focuses on responsible innovation, data security, transparency, and fair marketing practices to ensure that neurotechnological products are developed and deployed without compromising individual rights or safety. While societal ethics focuses on the collective, corporate ethics emphasizes accountability within business practices.

4.2.3 Corporate ethics in neurotechnology: a responsible framework

Corporate ethics in neurotechnology presents a multifaceted challenge that requires adherence to principles ensuring ethical business practices and the responsible development of emerging technologies such as BCIs. As neurotechnology companies navigate the intersection of technological innovation and human cognition, they must embed ethical considerations into every operational level. The principles of respect for persons, beneficence, non-maleficence, justice, equity, and solidarity serve as a foundation for the development of a responsible and ethical framework that safeguards the interests of all stakeholders, including users, employees, and society at large.

The principle of respect for persons demands that corporations uphold the dignity, privacy, and autonomy of all individuals affected by their business operations. In the neurotech industry, this extends to protecting sensitive neural data and ensuring that users fully control how their cognitive information is collected, processed, and utilized. Corporations must also respect the rights of employees, customers, and communities, implementing transparent data practices and maintaining rigorous standards for informed consent. In neurotechnology, this principle requires a heightened awareness of the ethical implications of manipulating or interfacing with the human brain, emphasizing the importance of trust, transparency, and accountability in all corporate dealings.

Beneficence in corporate ethics mandates that neurotechnology companies prioritize the well-being of their stakeholders by developing products that demonstrably improve health outcomes or enhance the quality of life for users. Neurotechnologies such as BCIs have the potential to profoundly impact cognitive and neurological functionality, which makes it essential that companies engage in research and development efforts that focus on therapeutic, rehabilitative, or quality-of-life improvements. Beneficence obliges corporations to go beyond profit motives to ensure that their innovations are designed to benefit individuals and society as a whole. This principle also drives corporations to pursue continuous product improvement, postmarket surveillance, and user feedback to optimize/maximize the positive impacts of neurotechnologies.

The principle of non-maleficence imposes an ethical obligation on corporations to avoid causing harm through their products, services, or business practices. This principle takes on particular significance in the context of neurotechnologies, as BCIs and other cognitive-enhancing devices directly affect neurological processes. Corporations must ensure that their products are rigorously tested for safety and efficacy, with robust safeguards to prevent physical, psychological, or social harm to users. Furthermore, data privacy and cybersecurity protections must be prioritized to prevent unauthorized access to or misuse of sensitive neurodata and maintain the integrity and trustworthiness of neurotechnological systems.

Justice within corporate ethics demands fairness and equity in all business practices, from labor rights and supplier relationships to customer interactions. In neurotechnology, this principle ensures that corporate activities do not intensify existing inequalities in healthcare or inequitable access to advanced technologies. Justice requires companies to make their products and services available to a broad demographic, avoiding practices that prioritize wealthier or more privileged consumers while neglecting underserved populations. Companies must engage in fair and transparent business practices, ensuring that the benefits of neurotechnological advancements are available to everyone.

Equity further expands the notion of fairness by addressing societal structural inequalities that may influence who benefits from neurotechnological innovations. Neurotechnology companies must actively work to reduce disparities in access, making concerted efforts to ensure that their products and services are affordable, available, and accessible to diverse populations. This principle compels corporations to mitigate social inequalities, especially in healthcare, by developing inclusive technologies that benefit marginalized or underserved communities.

Solidarity in corporate ethics encourages collaboration between industry peers, regulators, and the broader community to address the ethical implications of neurotechnology. No corporation can effectively navigate the ethical landscape of a field that is so highly complex and impactful. Solidarity fosters a cooperative approach, where companies work together to establish ethical standards, share best practices, and engage in a transparent dialogue with regulators, ensuring that the development and deployment of neurotechnologies are aligned with societal values. This principle promotes collective responsibility in advancing neurotechnological innovations while safeguarding ethical standards.

4.2.4 Applied ethics in neurotechnology: an ethical imperative

Applied ethics in neurotechnology demands a rigorous, principled approach to ensure that the development and application of BCIs and other cognitive technologies respect human rights and promote societal well-being. As neurotechnological advancements push the boundaries of human cognition and brain function, the integration of ethical principles into their design and deployment becomes a critical responsibility. The foundational principles of respect for persons, beneficence, non-maleficence, justice, and equity serve as a framework to guide the ethical application of these technologies, ensuring that their impacts are beneficial, inclusive, and just.

Respect for persons in the context of neurotechnology refers to the obligation to honor individual autonomy, privacy, and dignity. Given the intimate nature of neurotechnologies, which directly interface with neural processes, this principle demands stringent protocols relating to informed consent. Users must be fully aware of the potential risks, benefits, and long-term implications of engaging with neuro-technological systems. This user awareness is critical, since neurotechnologies generate and process vast amounts of personal cognitive data. Transparency is essential to safeguarding individual autonomy, which requires that users retain control over their neurodata and are free to opt in or out of such technologies without coercion. Furthermore, dynamic consent models, where consent evolves with technological updates, are critical in upholding respect for persons, given the iterative nature of progress in neurotech.

The principle of beneficence in applied ethics ensures that neurotechnological applications are explicitly designed to improve the cognitive, sensory, or motor abilities of users, thereby contributing to their well-being. In practice, this means that the development of BCIs and related technologies must be grounded in a user-centered approach, where the primary goal is to enhance the quality of life for individuals. Beneficence also emphasizes the need for continuous research and development to refine neurotechnologies to optimize/maximize their therapeutic and enhancement potentials. By focusing on measurable positive outcomes—such as improved cognitive function, better management of neurological disorders, or enhanced communication for individuals with disabilities—this principle ensures that neurotechnology serves a clear and beneficial purpose.

Non-maleficence, often understood as the principle of 'do no harm,' holds particular importance in neurotechnology due to the high stakes involved in manipulating brain function. This principle demands careful consideration of potential adverse effects, including short-term and long-term risks associated with neurotechnological interven-tions. Neurotechnologies must undergo rigorous preclinical testing and clinical trials to ensure their safety, efficacy, and reliability. Additionally, developers must implement robust cybersecurity measures to protect neurodata from unauthorized access or manipulation. By prioritizing the prevention of harm, non-maleficence safeguards users from unintended consequences that could arise from improperly designed or inad-equately tested neurotechnological applications.

The principle of justice in applied ethics addresses the equitable distribution of neurotechnological benefits across all groups in society. As these technologies become more prevalent, there is a risk of deepening existing social and economic inequalities if access is limited to only privileged populations. Justice mandates that neurotechno-logical innovations be made available to the entire spectrum of humanity, regardless of socioeconomic status, geographic location, or other barriers. This principle also highlights the need for regulatory frameworks and public policies to guarantee that neurotechnological advancements are not monopolized by a select few but integrated into public healthcare systems and made available to all.

Equity, which is closely related to justice, addresses specific structural inequalities that may prevent certain populations from benefiting from neurotechnologies. This principle requires proactive efforts to ensure that neurotechnologies are designed and implemented to accommodate diverse cognitive, cultural, and physical needs. Equity also involves reducing (or waiving) the cost of these technologies to make them accessible to underprivileged and marginalized communities. By actively working to dismantle barriers to access, the principle of equity ensures that neurotechnological advancements do not intensify social disparities but contribute to a more inclusive technological landscape.

4.2.5 Bioethics and medical ethics considerations

Both bioethics and medical ethics play critical roles in guiding the ethical integration of neurotechnology, particularly in the case of BCIs. However, their scope and focus diverge in meaningful ways, particularly regarding the broader societal impacts addressed in bioethics and the patient-centered approaches inherent in medical ethics. While closely related, both focus on ensuring ethical healthcare principles, particularly regarding patient autonomy, beneficence, non-maleficence, and justice. In neurotechnology, both fields aim to safeguard the rights of patients, prioritize their well-being, and ensure equitable access to advanced treatments such as BCIs.

However, the key difference lies in their scope. Medical ethics is more narrowly focused on clinical care, which emphasizes the direct responsibilities of healthcare providers in ensuring patient safety and informed consent in medical settings. In contrast, bioethics takes a broader approach, as it addresses the ethical implications of technologies that modify human biology, including their societal, legal, and long-term impacts. This distinction is critical for neurotechnology. Medical ethics govern the safe, responsible use of neurotech in patient care, while bioethics encompasses wider issues, such as how BCIs could reshape human cognition, privacy, and identity in nonclinical environments, raising broader ethical concerns.

In bioethics, respect for persons emphasizes the preservation of autonomy on a broader societal level to ensure that individuals maintain control over their cognitive data, particularly in nonclinical environments such as education or workplace settings. Neurotechnologies such as BCIs pose significant ethical challenges in maintaining autonomy, particularly regarding how data is collected, stored, and used. Bioethical frameworks demand dynamic, ongoing consent to ensure that individuals have the ability to revoke or modify their consent as these technologies evolve. In medical ethics, respect for persons focuses on patient autonomy in clinical settings, which highlights informed consent prior to any intervention. Patients must be fully aware of the risks, benefits, and long-term consequences of BCI interventions. The capacities of patients to make autonomous decisions regarding their treatments are paramount, which requires iterative consent to account for the rapid development of neurotechnological capabilities.

The bioethical perspective on beneficence necessitates that neurotechnologies akin to BCIs serve the public good by delivering tangible societal benefits without causing harm. Developers must ensure that these technologies enhance human capabilities and quality of life while minimizing risks through extensive testing and validation. In contrast, medical ethics applies beneficence within the context of individual patient care. The primary obligation is to ensure that BCI technologies provide the safest and most efficacious treatments. Healthcare professionals must carefully balance the potential therapeutic benefits of BCIs with their associated risks, while always prioritizing the well-being of patients.

Although both bioethics and its medical counterpart share the principle of non-maleficence, the bioethical view takes a broader approach focusing on preventing societal harm, ensuring that BCIs do not magnify social disparities or pose cybersecurity risks that might translate to the misuse of neurodata. In medical ethics, non-maleficence is more narrowly applied to individual patient safety. Healthcare providers must mitigate the cognitive, physical, and emotional risks that BCIs may bring about. Rigorous preclinical testing and continuous patient monitoring are essential to ensure that BCI technologies do not cause unintended harm during or following interventions.

The tenet of justice in bioethics embodies the fair distribution of BCI benefits and risks across diverse societal groups. It promotes strategies that ensure the accessibility of neurotechnologies for underserved and marginalized communities to prevent disparities in access to these innovations. In medical ethics, justice ensures that all patients, regardless of socioeconomic status, have equitable access to advanced neurotechnological treatments. Healthcare systems must provide these technologies to a wide spectrum of patients to prevent a widening gap in healthcare inequalities.

Bioethics supports equity by ensuring that neurotechnological advancements benefit all individuals, regardless of socioeconomic or social standing. It stresses the importance of inclusivity and affordability in designing and deploying BCIs to promote broader social justice. In medical ethics, equity aims to remove access barriers within healthcare systems to guarantee that innovative neurotechnologies are available to all, particularly underserved populations. The principle of solidarity in both fields emphasizes global cooperation and interdisciplinary collaboration to address neuroethical challenges and improve overall outcomes, whether for society or individual patients.

4.2.6 Brain–computer interface research trends

Recent BCI research reveals significant technological advances while highlighting the critical need for ethical frameworks to guide their development and implementation. This chapter provides a synthesis of critical publications (from 2021) that underscore the evolving capabilities of BCIs and their accompanying ethical considerations, followed by a selected list of further research (to 2025) in this important domain.

Krishnan and Bai (2021) focus on BCIs in multimedia environments and illustrate how these interfaces have the potential to transform user interactions with complex media systems. Their study, published in the *International Journal of Communication Systems*, emphasizes the urgent need for ethical guidelines to address user safety and data privacy concerns as BCIs become more integrated with digital environments.

Chicaiza and Benalcázar (2021) and Castillo and Delgado (2021) explore a BCI system that is designed to control Internet of Things (IoT) devices using electroencephalogram (EEG) signals. Their research, presented at the 2021 IEEE Fifth Ecuador Technical Chapters Meeting (ETCM), demonstrates the expanding role of BCIs in everyday technologies. The authors raise important ethical questions regarding the security and privacy of sensitive neurodata in personal and home settings.

Guger *et al* (2024) provide a comprehensive overview of BCI research in *Brain–Computer Interface Research: A State-of-the-Art Summary 11*. Their work highlights rapid advancements in BCI technologies while underscoring the pressing need for ethical frameworks to responsibly manage these developments, especially as BCIs become more integrated into human–computer interaction systems.

Addressing the technical challenges of BCIs, Serrano-Amenos *et al* (2024) present a simulation-informed power budget estimate for a fully implantable BCI in the *Annals of Biomedical Engineering*. Their research stresses the importance of sustainable and safe design practices in BCI development to ensure that future technologies are ethical and viable for long-term use.

Zhu *et al* (2021), writing in *IEEE Communications Surveys & Tutorials*, explore the potential for BCIs to enable a human-centric metaverse. Their research points to the transformative power of BCIs in virtual environments yet raises concerns about consent, privacy, and user well-being, which must be addressed through robust ethical guidelines.

King *et al* (2024) systematically review the risks associated with BCIs in the *International Journal of Human–Computer Interaction*. Their findings highlight the psychological and physical risks of BCI technologies and advocate for comprehensive ethical standards for their effective mitigation.

Saway *et al* (2024) discuss the evolution of neuromodulation for chronic stroke treatment, moving from neuroplasticity mechanisms to BCIs in neurotherapeutics. This research illustrates the therapeutic potential of BCIs while emphasizing the ethical implications, particularly regarding patient consent and the long-term effects of neurotechnological interventions.

Chai *et al* (2024) propose the use of BCI digital prescriptions for neurological disorders in *CNS Neuroscience and Therapeutics*. Their work raises critical ethical questions about the prescription and administration of digital therapies, including how they are monitored and regulated.

Qu *et al* (2024), in *Disability and Rehabilitation: Assistive Technology*, provide a meta-analysis of BCIs in post-stroke rehabilitation, focusing on upper-limb function. This research underscores the therapeutic benefits of BCIs while highlighting the need for ethical guidelines to ensure patient safety and treatment efficacy.

Finally, Meng *et al* (2024) address security concerns in BCIs, specifically adversarial filtering-based evasion and backdoor attacks in information fusion.

Their study underscores the vulnerabilities of BCIs to cyberattacks and emphasizes the necessity for stringent ethical standards to protect users from malicious actors.

Further selected recent research in this domain includes:

Bergeron *et al* 2023 (Use of invasive brain–computer interfaces in pediatric neurosurgery: technical and ethical considerations) *J. Child. Neurol.*

Ploesser *et al* 2024 (electrical and magnetic neuromodulation technologies and brain–computer interfaces: ethical considerations for enhancement of brain function in healthy people—a systematic scoping review) *Stereotact. Funct. Neurosurg.*

Rosenfeld 2024 (Neurosurgery and the brain–computer interface) *Adv. Exp. Med. Biol.*

Waisberg *et al* 2024 (Ethical considerations of neuralink and brain–computer interfaces) *Ann. Biomed. Eng.*

Zhang *et al* 2024 (Brain–computer interfaces: the innovative key to unlocking neurological conditions) *Int. J. Surg.*

Almanna *et al* 2025 (Public perception of the brain–computer interface based on a decade of data on X: mixed methods study) *JMIR Form. Res.*

Jaszczuk *et al* 2025 (Advances in neuromodulation and digital brain-spinal cord interfaces for spinal cord injury) *Int. J. Mol. Sci.*

Mehta 2025 (Brain–computer interface tool use and the contemplation conundrum: a blueprint of mental action, agency, and control) *Oxf. Open. Neurosci.*

Tzimourta 2025 (Human-centered design and development in digital health: approaches, challenges, and emerging trends) *Cureus.*

4.3 Section II: redefining digital identity in brain–computer interfaces

4.3.1 Redefining identity

In psychology and psychoanalysis, identity refers to an individual's self-concept, which is shaped by personal experiences, memories, relationships, and social roles. It involves the development of a cohesive sense of self that is influenced by unconscious processes, emotional states, and social interactions. In philosophy, identity is often explored in relation to existence, self-awareness, and the nature of being, which focuses on questions of continuity and change over time. In contrast, digital identity is a constructed representation of an individual in the digital realm, which is created through data such as social media profiles, online activities, and interactions across platforms. This digital self may synergize with one's psychological or psychoanalytic identity when it authentically represents their beliefs, emotions, and values. However, it can also clash with psychological identity, particularly when curated digital personas misalign with internal self-perceptions, leading to cognitive dissonance and identity fragmentation.

In the deep tech era, digital identity has evolved into a complex, multidimensional construct that spans numerous industries and platforms. Digital identity represents an individual's or entity's virtual presence, which is shaped by the aggregation of data and interactions across various digital services. With the proliferation of advanced technologies such as AI, blockchain, and the IoT, digital identity transcends its original forms and becomes a critical asset for access, personalization, and security across sectors. The growing importance of digital identity in this era requires a nuanced understanding of its categories and applications across industries, use cases, and platforms.

Personal identity forms the foundational layer of digital identity, which includes vital personal identifiers such as name, date of birth, social security number, and residential address. This identity is essential for accessing critical services, particularly in the government and healthcare sectors. Governments increasingly rely on digital identities for public services, from tax filings to national identification systems. Simultaneously, healthcare providers use digital identity for patient verification, access to medical records, and the secure delivery of telehealth services. In the deep tech era, personal identity is also central to privacy and cybersecurity frameworks, where verifying authenticity and protecting personal data from cyber threats are critical.

Professional identity encapsulates an individual's career history, educational background, professional skills, and certifications. In the business world, digital platforms such as LinkedIn and corporate databases utilize this identity for recruitment, networking, and career development. With AI-powered tools and blockchain-based credentials, one's digital identity in professional contexts now includes verified certifications and automated skills assessments, which enable a more efficient and secure labor market. A professional identity is increasingly intertwined with digital reputation systems, where blockchain and decentralized platforms play a role in establishing transparent, immutable records of one's work history and performance.

Educational identity is crucial for academic and lifelong learning ecosystems. This digital identity includes academic transcripts, certifications, coursework, and achievements within educational platforms. As e-learning platforms proliferate, especially in the digital transformation era in education, educational identities are increasingly used for credential verification, access to online courses, and academic collaborations. Blockchain technologies offer the potential for immutable educational credentials, allowing learners to carry verified records of their skills and achievements across platforms and institutions.

Social media identity reflects an individual's virtual presence on various platforms. It is built through social interactions, posts, and activities that create a digital footprint representing one's personal interests, connections, and social behaviors. In the deep tech era, social media identities are increasingly pivotal for personalized advertising, influencing consumer behavior, and even political campaigns. However, this also raises concerns about privacy, data ownership, and social platforms' ethical use of personal information.

Financial identity includes banking details, transaction histories, credit scores, and financial assets. Financial institutions, e-commerce platforms, and fintech services use it for secure transactions, lending decisions, and financial monitoring.

Blockchain and AI are revolutionizing financial identity by enabling more secure, decentralized financial services and automating financial decision-making processes. Digital wallets, decentralized finance (DeFi), and cryptocurrency also increasingly reshape how financial identity is managed and verified.

Health identity encompasses medical records, biometric data, health history, and prescription information, which are critical for accessing healthcare services and interacting with health monitoring apps. With the advent of telemedicine, wearable health devices, and AI-driven diagnostics, health identity is becoming increasingly digital. This shift highlights the need for robust data security and privacy regulations, given the sensitivity of health data and its growing role in personalized medicine and healthcare delivery.

Finally, consumer identity involves an individual's shopping behaviors, preferences, loyalty programs, and online transaction history. Retailers and marketing platforms use this identity to tailor customer experiences, offering personalized recommendations, promotions, and loyalty rewards. AI and machine learning algorithms enhance one's consumer identity by predicting purchasing patterns and enabling hyper-personalized marketing while raising concerns about consumer privacy and data ownership.

4.3.2 Multiple digital identities

The digital era is characterized by the seamless integration of physical and digital experiences, where technology bridges real-world interactions with digital platforms. This convergence has led to multiple digital identities, as individuals manage distinct personas across professional, social, educational, and financial domains in the digital space. Individuals must manage multiple digital identities for different personal, professional, and financial purposes. While serving specific functions, these identities offer distinct benefits and significant risks. From enhancing personalization to greater privacy and security concerns, the management of multiple digital identities has become increasingly complex as the digital and physical worlds converge.

While the benefits of managing multiple digital identities are clear, the risks associated with this complexity are equally significant. One major benefit is the ability to tailor each digital identity to its context, enhancing personalization and optimizing user experiences. For example, an individual's educational identity can assist with personalizing learning pathways, while their social media identity can be used to curate content that is aligned with their personal interests. The convergence of multiple identities allows for more integrated and intuitive interactions across different platforms, creating a unified experience that reflects diverse aspects of life.

However, managing multiple digital identities presents serious privacy and security challenges. As users engage with various platforms, they expose different facets of their personal information, which makes it easier for third parties or malicious actors to create comprehensive personal profiles. The fragmentation of digital identities across multiple services also increases the risk of data breaches and identity theft. Cybercriminals can exploit vulnerabilities in one identity (such as weak passwords or insufficient encryption) to access other sensitive information, potentially compromising financial, educational, or professional data.

The ascent of digital identities has complex implications for individuals, corporations, and society. While digital identities create opportunities for personalization and efficiency, they also introduce significant challenges that surround data privacy, security, and ethical use. It is essential to distinguish between digital ethics and cybersecurity in the context of digital identity, as they address distinct but interconnected concerns. Digital ethics focuses on the responsible use of digital identities, whereas cybersecurity aims to protect these identities from malicious threats such as data breaches, fraud, and cybercrime.

Managing multiple digital identities across platforms such as social media, banking, healthcare, and e-commerce is empowering, albeit risky for individuals. These identities include personal, professional, and financial data, making them valuable targets for cybercriminals. Cybersecurity strategies, such as encryption, multifactor authentication, and regular monitoring, are critical for protecting these digital identities from unauthorized access and breaches. However, digital ethics goes beyond protection, ensuring that personal data is handled with respect for autonomy, privacy, and informed consent. Ethical considerations arise when digital identities are commodified, with individuals having limited control over how corporations, governments, or third parties use their data. The clash between data ownership and user rights often leads to ethical concerns that cannot be fully addressed by cybersecurity measures alone.

Corporations benefit immensely from digital identities by gaining insights into customer preferences, behaviors, and needs, which enable them to offer more personalized products and services. However, the corporate responsibility for digital identity management extends beyond mere compliance with data protection laws such as the General Data Protection Regulation (GDPR) (Lawton 2020) and the California Consumer Privacy Act (CCPA) (Nguyen 2022). While cybersecurity measures protect sensitive customer data, digital ethics compels companies to consider how they collect, use, and store this data. For instance, ethical concerns arise when companies exploit user data for profit without explicit consent or transparency, which often leads to breaches of trust. The management of ethical digital identity requires corporations to be transparent in terms of their data practices to ensure that users are fully informed and have control over their digital footprints. This distinction highlights how digital ethics encompasses more than just preventing data breaches; it also addresses the broader implications of how digital identities are used and shared.

In human resource (HR) management, digital identities are critical for everything from recruitment to performance evaluation. Cybersecurity ensures that employee data (personal information, credentials, and professional evaluations) remains protected from external threats. However, digital ethics examines how companies use this data internally. For example, algorithms that assess the digital identities of employees may introduce bias or unfairly influence hiring decisions, which raises ethical concerns regarding transparency and accountability in data-driven HR processes.

At the societal level, the proliferation of digital identities highlights the importance of access and inclusivity. Governments and regulatory bodies must create frameworks that address cybersecurity risks and the ethical use of digital identities to achieve social equity. While cybersecurity protects the technical aspects of identity management (e.g. securing digital infrastructure and preventing cybercrime), digital ethics addresses broader concerns such as data sovereignty, surveillance, and the ethical implications of digital exclusion. The standardization of digital identity protocols and the creation of interoperable systems are essential to ensure that all individuals, regardless of socioeconomic status, can effectively manage their digital identities.

4.3.3 Digital identity in brain–computer interfaces

Digital identity takes on an expanded and more complex meaning for BCIs. Traditionally, digital identity referred to the virtual representation of a person across platforms, which encompassed their social, financial, professional, and personal personas. For BCIs, this concept is extended to include the unique digital representation of an individual's neurological data and cognitive processes. BCIs facilitate direct communications between the brain and digital systems, which results in a dynamic, intricate digital identity that reflects an individual's mental and cognitive state. BCIs collect and analyze vast quantities of neurological data, transforming brain signals into actionable outputs in digital environments. These interactions create unique digital profiles that integrate with the broader digital identities of users, combining the traditional facets of personal, professional, educational, and social identities with real-time cognitive and neurological data. The fusion of neurological data with conventional digital identities opens up new possibilities while presenting unique ethical, privacy, and security challenges.

Educational identity in the context of BCIs might involve enhancing learning experiences through neurofeedback systems that adapt to the cognitive state of the user. For instance, BCIs integrated into e-learning platforms might monitor a student's brain activity during coursework and adjust the delivery of content in real time, creating a personalized learning experience based on their engagement, focus, or cognitive load.

Professional identity and BCIs intersect in the workplace, where BCIs could optimize productivity by tracking cognitive states such as attention, fatigue, or stress. A professional identity enhanced by BCI data might be used for real-time performance evaluations, adaptive workload management, or cognitive training programs. For example, an employee's brain activity could be monitored to assess focus levels, and tasks might be dynamically adjusted to match cognitive readiness. This integration of BCI data into professional identity raises critical ethical questions regarding workplace surveillance, data ownership, and the potential for the misuse of cognitive data by employers.

BCIs may also dramatically transform one's social media identity, as their potential to monitor emotional states and cognitive responses introduces new ways of interacting with social platforms. For instance, social media could evolve to include textual or visual posts and emotional feedback loops that are directly derived from brain signals, enabling users to communicate in more immersive, neuroenhanced ways. However, this also introduces risks, such as the exploitation of emotional and cognitive data for marketing or behavioral manipulation, which raises significant privacy and consent concerns.

Financial identity could similarly be affected by the integration of BCIs. In financial services, BCIs could facilitate secure transactions through biometric brainwave authentication, reducing the need for traditional passwords or PINs. This would create more seamless interactions between users and financial systems while amplifying the need for cybersecurity measures to protect against unauthorized access to brainwave data. The combination of financial history and real-time neural data introduces new dimensions to identity theft, making it critical to develop robust safeguards.

The continuous and dynamic nature of BCI data introduces unique risks in terms of privacy and security. Traditional digital identities are already vulnerable to data breaches and misuse; however, BCI-enhanced identities include even more sensitive information, such as real-time brain activity patterns. Therefore, digital identity management in the context of BCIs is more complex and demands far higher levels of security to protect users from unauthorized access or the exploitation of their neurological data. Cybercriminals could target BCI systems to manipulate or steal cognitive data, leading to profound consequences for one's personal, professional, and financial identities.

Additional ethical dilemmas can arise as BCI-enhanced digital identities become more integrated into daily life. The collection and use of neurological data introduces questions of autonomy, consent, and data ownership. Who owns the brain data collected by BCIs? How can individuals ensure that their cognitive profiles are used ethically and not commercially exploited? These questions must be addressed through clear regulatory frameworks that prioritize user rights while ensuring that risks to privacy and autonomy do not outweigh the benefits of BCI technologies.

4.3.4 Dynamic informed consent in brain–computer interfaces

In the digital era, informed consent has evolved into a complex, multilayered process. Ideally, individuals should be fully aware of how their personal data is collected, used, and shared before they engage with digital platforms or services. Digital informed consent demands that users clearly understand how their data will be processed, who will have access to it, and the potential risks involved. It emphasizes user autonomy, ensuring that individuals voluntarily agree to data collection and usage terms after receiving comprehensive and understandable information. Users are now required to consent to the initial data collection and the long-term implications of how their data could be utilized, shared, or even sold to third parties.

For BCIs, informed consent takes on heightened importance due to the sensitive nature of the data involved. Brain data is profoundly personal, as it encompasses thoughts, emotions, and cognitive processes beyond conventional digital data such as browsing histories or social media activities. Users must understand how the system works, the type of data being collected, and the potential risks, including security breaches, misuse of data, or unauthorized access by third parties. Further, because BCIs would involve continuous data collection, the need for dynamic user consent is imperative. Unlike traditional forms of digital data, which may be retrieved at specific times, BCIs create a constant flow of sensitive neurological information, necessitating the management of continuous consent.

4.4 Section III: ethical considerations in hybrid augmented brain–computer interfaces

4.4.1 Ethics of hybrid augmented workflows: navigating complex human–digital interactions

The ethical management of hybrid augmented workflows that involve elaborate interactions between humans and digital systems is becoming increasingly crucial as technology advances. These workflows, which integrate human decision-making with sophisticated digital tools, span various domains, including AI, blockchain, digital twins, and BCI-powered systems. The ethical challenges of these systems are complex, particularly in cases where technologies (e.g. BCIs) directly interact with human neural processes. This chapter explores the ethical considerations associated with human-in-the-loop workflows across these domains, which emphasize the unique challenges posed by BCIs.

4.4.1.1 Human-in-the-loop workflows and ethical implications

In AI-enhanced BCI workflows, humans often interact with machine learning models to refine and validate outcomes. For example, radiologists use AI to assist with the detection of anomalies in medical imaging. However, ethical concerns arise regarding the transparency of AI decision-making processes, potential biases in model training, and the need to maintain human oversight to prevent an over-reliance on automated systems. Furthermore, privacy concerns are significant, given the vast amount of data required to train AI models. Ensuring the ethical use of AI requires robust frameworks that prioritize fairness, transparency, and human control over decision-making processes.

4.4.1.2 Unique ethical challenges of hybrid augmented intelligence powered by a brain–computer interface

BCIs represent a significant leap in human–digital interactions, directly linking human neural activities with digital systems. This introduces profound ethical challenges, particularly concerning mental privacy, cognitive autonomy, and the potential for manipulating human thought processes. To navigate these unique issues, BCIs require a deep understanding of neuroscience, human executive function, and cybersecurity. One of the primary concerns for BCIs involves their impacts on neuroscience and human executive functions. BCIs collect and interpret neural signals, which raises important ethical questions regarding mental privacy and cognitive autonomy. The capacity of BCIs to potentially influence or modify thought processes necessitates stringent ethical guidelines to protect individuals from manipulation. Any enhancements or interventions facilitated by BCIs must be consensual, transparent, and beneficial to the individual.

Privacy and data security concerns are particularly pronounced in BCI systems, as they deal with the most intimate form of data—neural activity. Unauthorized access to or manipulation of this data could lead to unprecedented privacy violations and psychological harm. Protecting the integrity of neural data requires robust cybersecurity measures to guard against cyberattacks that might threaten an individual's mental health and autonomy. Due to the invasive nature of BCI technologies, informed consent processes are especially vital. Individuals must thoroughly comprehend the potential risks and benefits associated with BCIs, as well as the long-term implications for their cognitive functions and autonomy. Unlike AI or blockchain, the direct impacts of BCIs on cognitive processes require more rigorous consent mechanisms to ensure that participants retain control over their neural data and decision-making capacities.

Finally, bias and accessibility in developing and deploying BCI technologies must be carefully considered to avoid exacerbating existing social inequities. Ethical frameworks must ensure that these advanced technologies are accessible to all, regardless of socioeconomic status, and that the benefits of BCIs are distributed equitably across society. Guaranteeing fairness and inclusivity in deploying BCIs is essential to prevent the creation of a privileged class of individuals who disproportionately benefit from these innovations.

4.4.1.3 Key features of hybrid augmented neurotechnology workflows

Hybrid augmented neurotechnology workflows represent the intersection between human cognition and advanced neurotechnologies; at this intersection, human cognitive processes are integrated with cutting-edge tools to enhance learning, decision-making, and behavioral adaptations. These systems offer unprecedented opportunities for the personalized augmentation of human capabilities. However, their implementation must be carefully designed to accommodate neurodiversity, ensure inclusivity, and maintain user autonomy. This chapter outlines the key features of hybrid augmented neurotech workflows, with an emphasis on ethical considerations and responsible deployment.

An essential ethical consideration in hybrid neurotechnology workflows relates to respect for neurodiversity. Human cognition is highly variable, as individuals have distinct modes of thinking, learning, and processing information. Ethical frameworks that govern neurotechnologies must prioritize this diversity to ensure that these workflows are adaptable and supportive of individual cognitive strengths. Rather than imposing a one-size-fits-all solution, neurotech systems must be flexible, allowing users to customize their interactions based on their unique cognitive profiles. This allowance for neurodiversity enhances individual performance and ensures that these technologies support inclusivity and accessibility across diverse populations.

Neurotechnology tools must adhere to universal design principles to follow ethical principles and attain optimal technology integration. Inclusive design guarantees that neurotech workflows are accessible to users regardless of their cognitive, sensory, or physical capacities. The concept of universal design promotes the creation of systems that provide identical or equivalent means of use for all individuals to ensure equitable access. By adopting such principles, neurotech workflows can bridge the gaps between diverse abilities to provide seamless interactions with neurotechnologies. This approach is essential for fostering an inclusive technological environment where all users, regardless of ability, can benefit from neurotechnology advancements.

Maintaining user control and autonomy is fundamental to the ethical deployment of neurotechnology workflows. Users should be able to customize their interactions with neurotech systems and exercise control over the functionalities they engage with while ensuring that these systems align with their cognitive preferences. This neurocentric design provides users with straightforward mechanisms to modify their preferences or opt out of certain features at any stage of their interactions. Autonomy is particularly critical when dealing with technologies such as BCIs, which directly interact with neural processes. Respecting individual agency in neurotech workflows reinforces trust and supports ethical principles of informed consent and control.

BCI-powered workflows must be tailored to accommodate human learning styles, intelligence types, and executive functions to optimize their effectiveness. BCIs facilitate direct interactions between neural processes and digital systems to potentially significantly augment human performance. To harness this potential, workflows must be designed to support diverse learning preferences (e.g. visual, auditory, and kinesthetic) to fully leverage individual interests and strengths. Furthermore, understanding and integrating different kinds of intelligence, such as analytical, creative, and emotional, can enhance the effectiveness of BCIs in augmenting human capabilities. These systems can improve decision-making and task management by aligning BCI applications with human executive functions such as memory, attention, and problem-solving skills, fostering a personalized and efficient user experience.

BCIs possess the transformative potential to reshape how we learn, process information, and interact with digital systems by directly linking human cognitive processes to external devices. To maximize their potential while upholding ethical principles, they must be designed to respect individual autonomy, enhance educational outcomes, and ensure inclusivity across diverse populations. By tailoring BCIs to various learning styles and intelligence types, we can create educational tools that are effective, equitable, and respectful of the cognitive diversity of learners.

4.4.2 Tailoring brain–computer interfaces to diverse learning styles and intelligence

BCIs might significantly enhance learning by adapting to different intelligence types, such as linguistic, logical–mathematical, musical, and kinesthetic, as each requires unique approaches. For instance, individuals with high linguistic intelligence might use BCIs to boost language acquisition, while those with kinesthetic intelligence may benefit from improved motor coordination, especially for rehabilitation. BCIs may also enhance interpersonal communication by interpreting emotional cues or assist intrapersonal learners by offering insights into emotional regulation and stress management. By personalizing the learning experience to cognitive strengths, BCIs offer a promising tool for inclusive education that maximizes individual potential. However, implementing BCIs in education will need to follow ethical guidelines that prioritize autonomy, fairness, and equitable access. This technology must not only enhance educational outcomes but also respect the diverse needs of learners. By customizing BCIs to individual learning styles, these tools could dramatically transform education while promoting inclusivity and respect for cognitive diversity.

BCIs may also improve executive functionality, such as focus, emotional control, and working memory. They could monitor neural activities to detect distractions and provide timely interventions to maintain focus, enhancing productivity and learning. BCIs might adjust task complexity based on the mental loads of individuals who struggle with cognitive flexibility to ease transitions during challenging tasks. By tracking the neural signals associated with emotional states, BCIs could promote improved emotional regulation. Additionally, they could offer task prompts to over-come inertia and manage working memory by storing information, helping users organize tasks, and improving efficacy.

4.4.3 Impact of brain–computer interfaces on employment

In the employment context, BCIs must be governed by ethical principles to ensure positive outcomes for both employees and organizations. Respect for persons emphasizes the preservation of employee autonomy and securing informed consent when implementing BCIs at work. Beneficence promotes the use of BCIs to enhance job satisfaction and productivity while safeguarding employee well-being. Non-maleficence ensures that BCIs do not cause harm, such as increased stress or privacy violations, while justice mandates equal access to the benefits of BCIs to ensure that no employee is excluded based on role or status. Equity emphasizes the provision of accommodations and support so that employees of all abilities can fully benefit from BCI-enhanced tools. Finally, solidarity fosters a collaborative work culture where BCIs are utilized to promote collective welfare, team collaboration, and shared problem-solving to make certain that these technologies enhance rather than undermine the workplace environment.

4.5 Section IV: brain–computer interface harmonization

4.5.1 Brain–computer interface cyberethics

The development and deployment of BCI technologies raise significant ethical challenges that require robust governance frameworks. The ethical governance of BCIs must prioritize the protection of personal data, user autonomy, and the equitable distribution of associated benefits and risks. As BCIs evolve, verifying that they are inclusive, secure, and ethically sound requires adherence to crucial ethical principles: respect for persons, beneficence, non-maleficence, justice, equity, and solidarity. Central to this effort is the establishment of cyberethics frameworks that guide the secure and sustainable implementation of BCIs. The design of proactive cyberethics security and resilient ecosystems for BCIs is critical for safeguarding the rights and well-being of users. As BCIs directly interface with neural data (some of the most sensitive forms of personal information), it is imperative to establish ethical guidelines that prioritize transparency, security, and inclusivity.

4.5.2 Sustainable brain–computer interfaces

BCI sustainability must be a key focus, as the long-term environmental and social consequences of these technologies could significantly impact future generations. Respect for persons requires an acknowledgment of the lasting impact of BCI technologies on both current and future users. This principle compels developers to adopt sustainable practices that do not compromise the well-being of future generations, for example, by minimizing resource depletion and any environmental harm.

4.5.3 Diverse and inclusive brain–computer interfaces

Beyond sustainability, the ethical deployment of BCIs must prioritize inclusivity and diversity. Respect for persons underscores the importance of recognizing and valuing individual differences in the design and application of BCIs. This principle ensures that diverse perspectives and experiences are not just considered but actively sought throughout the innovation process, making BCIs accessible and effective for a wide range of users.

4.6 Section V: deep tech convergence

The integration of BCIs with emerging technologies such as AI, digital twins, multiomics, multicloud computing, high-speed networks, satellite internet, and quantum computing represents a transformative shift in how humans engage with technologies. These systems can potentially revolutionize learning and interactivity by creating highly personalized and responsive environments. However, their ethical deployment requires a careful, nuanced approach that considers diverse human learning styles and intelligences while remaining rooted in a deep understanding of neuroscience, human cognitive functions, and ethical principles.

4.6.1 Brain–computer interfaces and AI

Combining BCIs with a complex AI portfolio gives rise to exciting opportunities and challenges, revolutionizing human–computer interactions and cognitive enhancements. Integrating BCIs with generative AI might enable the real-time creation of personalized content based on neural feedback loops. For instance, creative professionals might generate artwork, music, or written content by imagining concepts and expanding creative possibilities. This fusion may also streamline complex processes, allowing individuals to design solutions or innovations more rapidly and intuitively. Pairing BCIs with cognitive AI (which mimics human thought processes) could enhance decision-making and learning. BCIs might help individuals access cognitive AI systems that anticipate user needs based on neural signals, tailoring learning materials or decision-support systems to the user's unique cognitive state. Adaptive AI could further enhance personalization by learning from the user's mental state and dynamically adapting. When integrated with BCIs, adaptive AI might adjust tasks, provide tailored interventions when users are under stress or cognitive overload, and improve real-time task management to optimize productivity and well-being.

Neuromorphic AI, inspired by the structure and functionality of the human brain, benefits significantly from BCI integration. This technology can process sensory data that aligns with natural neural processing and offers intuitive interactions, particularly in domains such as music or kinesthetic learning. For instance, neuromorphic AI might customize music learning experiences to a user's emotional responses (detected via BCI) to create an instinctive and engaging environment. This synergy between BCIs and neuromorphic AI would foster a learning experience that is in harmony with the user's cognitive and sensory patterns.

However, these integrations bring significant ethical and privacy concerns, as neural data is highly sensitive, and the security of brain data and privacy is paramount. Unauthorized access or the misuse of this data could lead to nefarious manipulation or cognitive exploitation. Another challenge is user autonomy. Adaptive and cognitive AI systems may over-rely on real-time neural inputs, which raises concerns about manipulation and loss of personal agency if the system influences decision-making without full user control. BCI integration also demands substantial computational power and seamless interactions between neural inputs and AI systems, which present technical processing, latency, and accuracy hurdles.

4.6.2 Brain–computer interfaces and digital twins

Digital twins, which are virtual replicas of physical entities used for simulations, can be significantly enhanced by BCIs. By enabling direct cognitive interactions with these virtual models, BCIs may allow individuals to manipulate and explore digital twins using their thoughts alone. For example, individuals with high spatial intelligence might use a BCI to navigate and modify architectural models and intuitively engage with virtual environments. This combination of BCIs and digital twins can enrich the learning process, providing a platform that adapts to individual learning preferences, particularly for those who thrive on visual and spatial interactions. This collaborative cognitive engagement promotes a deeper, more meaningful learning experience.

4.6.3 Precision brain–computer interfaces

Future research and development in precision BCIs powered by multiomics will likely focus on the integration of genomic, proteomic, metabolomic, and epigenomic data to customize neurotechnological interventions to individual biological profiles. This approach might revolutionize personalized neurotherapies by optimizing BCI performance for neurological disorders, cognitive enhancements, and mental health treatments. By leveraging multiomics, BCIs could achieve unprecedented precision in understanding the complex molecular underpinnings of brain activities, enabling more effective, customized interventions. This fusion of multiomics with BCI technology will drive innovations in personalized medicine while advancing both neural rehabilitation and cognitive augmentation based on individual molecular landscapes.

4.6.4 Brain–computer interfaces and multicloud computing

The integration of BCIs with multicloud computing offers transformative potential by enabling the seamless, real-time processing of vast quantities of neural data across distributed cloud environments. Multicloud computing allows BCIs to leverage the strengths of different cloud platforms, enhancing scalability, resilience, and data storage capabilities. This integration provides the high-speed data transmission and processing that are required for sophisticated BCI applications, such as cognitive augmentation and personalized healthcare solutions. Furthermore, multicloud architectures can enhance data security and privacy by enabling more robust encryption and redundancy strategies, which are critical for safeguarding sensitive neurological data in a highly interconnected digital ecosystem.

4.6.5 Brain–computer interfaces and high-speed networks

Integrating BCIs with 5G and 6G networks could revolutionize real-time data transfer, as it would enable seamless communications between the human brain and digital systems. Ultralow latency and high-speed connectivity will enhance remote neural applications, from telemedicine to immersive augmented and virtual reality experiences.

4.6.6 Brain–computer interfaces and satellite internet

BCIs paired with satellite internet could expand neural connectivity across remote and underserved areas. This would facilitate global access to neurotechnologies, enabling real-time brain data transmission and empowering remote cognitive healthcare, education, and human–machine collaborations worldwide.

4.6.7 Brain–computer interfaces and quantum computing

The future of BCI development lies in its potential augmentation with quantum computing, which will unlock unprecedented computational power for the processing of complex neural data. Quantum computing could significantly enhance the capabilities of BCIs to interpret vast quantities of brain signals in real time, leading to more accurate and personalized interactions. This synergy may accelerate advancements in cognitive enhancement, neuroprosthetics, and mental health therapies. Additionally, quantum-enhanced BCIs might offer more secure data encryption to address privacy concerns. As quantum computing matures, its integration with BCIs could revolutionize fields such as neurotechnology, healthcare, and human–computer interactions, pushing the boundaries of human augmentation.

4.7 Section VI: strategic considerations

4.7.1 Corporate board-level considerations

At the corporate board level, BCI deployments must align with long-term organizational goals, compliance requirements, and ethical standards. The board should ensure that BCIs are integrated into business strategies, focusing on innovation, competitive advantages, and adherence to regulatory frameworks. Ethical guidelines surrounding data privacy, security, and user consent must be prioritized to safeguard company reputations while ensuring compliance with data protection laws. The board should also assess the societal and legal implications of using BCIs to ensure responsible and equitable technology deployment.

4.7.2 C-suite responsibilities

C-suite executives will play crucial roles in aligning BCI deployments with their organization's vision and operational strategies. They should focus on how BCIs may enhance productivity, innovation, and customer engagement. Equally important are their roles in fostering a culture of ethical awareness. This ensures that employees and customers understand and accept the benefits and potential risks of BCI usage. Leaders in the C-suite must also drive the development of solid cybersecurity measures to protect sensitive neurological data while anticipating market trends and regulatory changes to achieve a sustained competitive advantage.

4.7.3 Management-level responsibilities

Management teams are tasked with implementing BCIs effectively in day-to-day operations. This includes ensuring that the BCI technologies used are safe, functional, and aligned with organizational ethics. A key aspect of this is the provision of appropriate training for employees to ensure that BCI technologies are employed effectively and safely. Managers must also enforce data privacy policies and monitor the operational impacts of BCIs, addressing concerns related to privacy, performance metrics, and employee autonomy.

4.7.4 Employee-level considerations

At the employee level, BCIs must be introduced with transparency regarding their roles in enhancing productivity and well-being. Employees must have clear guidelines about how their cognitive data will be used, along with robust consent mechanisms. Their autonomy and privacy must be respected to verify that BCIs are used ethically and transparently to enhance job performance without compromising personal freedoms.

4.8 Future directions

The complexities of BCI technologies demand a multidisciplinary approach. The convergence of neuroscience and technology via BCIs requires collaboration across diverse fields, including ethics, engineering, law, and public policy. Experts from these disciplines must work together to navigate the moral, societal, and regulatory challenges that arise with the advancement of BCIs. This collaborative approach is crucial for ensuring that BCI innovations are at the forefront of technological progress and adhere to the highest standards of ethical and social responsibility. As BCIs advance toward broader adoption, they stand at the vanguard of a technological revolution that will seamlessly merge human intelligence with digital systems. While their potential to enhance human capabilities and transform our interactions with technology is immense, they must be harnessed within a framework of ethical vigilance. The responsible deployment of BCIs requires upholding human rights, safeguarding cognitive autonomy, and promoting social equity. As we venture into this new frontier of neurotechnology, we must be guided by a moral imperative to ensure that BCIs serve humanity's best interests, advancing progress without compromising ethical integrity.

The rapid evolution of immersive realities such as the metaverse, Industrial Omniverse™, and smart cities calls for BCIs that are tailored to these unique environments. As digital interactions become more integrated into daily life, BCIs must evolve to seamlessly connect human cognition with virtual and augmented realities to enable real-time neural interactions. In the metaverse, BCIs could offer unprecedented levels of immersion, facilitating intuitive controls, communication, and collaboration. In smart cities, BCIs might empower citizens by optimizing interactions with AI-driven infrastructures, enhancing mobility, communications, and decision-making. Future research and development must focus on designing BCIs that adapt to these complex environments to ensure that they align with cognitive diversity, privacy, and security requirements while fostering equitable access across diverse populations.

4.9 Conclusions: the moral imperative of responsible brain–computer interface deployment

As we stand on the threshold of an unprecedented technological epoch, BCIs signify a transformative leap in the integration of human neural processes with advanced digital systems. These cutting-edge technologies, positioned at the confluence of neuroscience, digital innovation, and human cognition, represent a pioneering frontier in the realm of exponential and fusing technologies. BCIs exemplify the synergistic melding of diverse domains (e.g. neuroscience, human–computer interaction, and artificial intelligence) into a unified framework poised to revolutionize the way we interface with the digital world.

The transformative capacities of BCIs are unparalleled in their potential to decode and harness neural activities for cognitive enhancement, physical augmentation, and novel forms of communication and control. By directly linking brain signals with digital platforms, BCIs can offer profound advances in areas such as neuroprosthetics, cognitive rehabilitation, and human–computer interactions. However, this immense potential is accompanied by equally profound ethical responsibilities. The development and deployment of BCIs must be meticulously governed by a robust ethical framework to ensure that their benefits are realized in ways that are equitable, inclusive, and respectful of human dignity.

BCIs stand at the intersection of technology and human consciousness, a convergence that offers both groundbreaking opportunities and significant ethical challenges. The integration of such powerful technologies into human life and society raises critical questions regarding cognitive autonomy, privacy, and the nature of human experience. As BCIs evolve, they will not only reshape our interactions with digital systems but also redefine the fundamental aspects of identity, agency, and personal control over one's neural data. Thus, it is imperative that the deployment of BCIs is grounded in comprehensive ethical principles, such as respect for persons, beneficence, non-maleficence, justice, equity, and solidarity.

The ethical imperative that will guide the trajectory of BCI development cannot be overstated. Respect for persons demands that individuals retain control over their cognitive processes and neurodata, ensuring that autonomy is preserved in the face of this invasive (even if not physiologically so) technology. Beneficence and non-maleficence obligate developers to design BCIs that enhance well-being while minimizing harm, particularly harm related to the potential psychological and social consequences of neurotechnologies. Justice and equity are essential for preventing BCIs from exacerbating existing societal inequalities or creating new digital divides. Solidarity calls for collective responsibility and cooperation to ensure that the advancements of BCIs benefit all of humanity rather than a select few.

The complexities of BCI technologies demand a multidisciplinary approach. The convergence of neuroscience and technology within BCIs requires collaborative efforts across diverse fields, encompassing ethics, engineering, law, and public policy. Experts from these disciplines must work together to navigate the moral, societal, and regulatory challenges that will inevitably arise. This collaborative approach is critical for ensuring that BCI innovations are not only at the forefront of technological progress but also adhere to the highest standards of ethical and social responsibility. As BCIs advance toward broader implementation, as stated above, they stand as the harbingers of a technological revolution that may ultimately seamlessly blend human intelligence with digital systems. While their potential to enhance human capabilities and transform our interactions with technologies is immense, they must be harnessed within a framework of ethical vigilance. The responsible deployment of BCIs requires a strong commitment to upholding human rights, safeguarding cognitive autonomy, and promoting social equity. As we venture into this new frontier of neurotechnology, we must be guided by a moral imperative to ensure that BCIs serve the best interests of humanity, advancing progress without compromising ethical integrity.

References and further reading

Almanna M A, Elkaim L M, Alvi M A, Levett J J, Li B, Mamdani M, Al-Omran M and Alotaibi N M 2025 Public perception of the brain–computer interface based on a decade of data on *X*: mixed methods study *JMIR Form. Res.* **9** e60859

Bergeron D, Iorio-Morin C, Bonizzato M, Lajoie G, Orr Gaucher N, Racine É and Weil A G 2023 Use of invasive brain–computer interfaces in pediatric neurosurgery: technical and ethical considerations *J. Child Neurol.* **38** 223–38

Castillo R J C and Delgado J M 2021 A brain–computer interface for controlling IoT devices using EEG signals *2021 IEEE Fifth Ecuador Technical Chapters Meeting (ETCM)* 1–6

Chai X, Cao T, He Q, Wang N, Zhang X, Shan X *et al* 2024 Brain–computer interface digital prescription for neurological disorders *CNS Neurosci. Ther.* **30** e14615

Chicaiza K O and Benalcázar M E 2021 A brain–computer interface for controlling IoT devices using EEG signals *2021 IEEE Fifth Ecuador Technical Chapters Meeting (ETCM)* 1–6

Guger C, Ince N F, Korostenskaja M and Allison B Z 2024 *Brain–Computer Interface Research: A State-of-the-Art Summary 11* ed C Guger, B Allison, T M Rutkowski and M Korostenskaja (Berlin: Springer)

Jaszczuk P, Bratelj D, Capone C, Rudnick M, Pötzel T, Verma R K and Fiechter M 2025 Advances in neuromodulation and digital brain-spinal cord interfaces for spinal cord injury *Int. J. Mol. Sci.* **26** 6021

King B J, Read G J M and Salmon P M 2024 The risks associated with the use of brain–computer interfaces: a systematic review *Int. J. Hum.-Comput. Interact.* **40** 131–48

Krishnan A and Bai V T 2021 Investigation of brain computer interface for rich multimedia environment *Int. J. Commun. Syst.* **34** e4584

Lawton A 2020 General data protection regulation: what does this mean in terms of law and ethics? *Arch. Dis. Child Educ. Pract. Ed.* **105** 294–5

Martins N R B *et al* 2019 Human brain/cloud interface *Front. Neurosci.* **13** 112

Mehta D 2025 Brain–computer interface tool use and the contemplation conundrum: a blueprint of mental action, agency, and control *Oxf. Open Neurosci.* **4** kvaf002

Meng L, Jiang X, Che X, Liu W, Luo H and Wu D 2024 Adversarial filtering based evasion and backdoor attacks to EEG-based brain–computer interfaces *Information Fusion* **107** 102316

Nguyen T 2022 An empirical evaluation of the implementation of the California Consumer Privacy Act (CCPA) *ArXiv* https://arxiv.org/abs/2205.09897 (accessed 18 July 2025)

Ploesser M, Abraham M E, Broekman M L D, Zincke M T, Beach C A, Urban N B and Ben-Haim S 2024 Electrical and magnetic neuromodulation technologies and brain-computer interfaces: ethical considerations for enhancement of brain function in healthy people—a systematic scoping review *Stereotact. Funct. Neurosurg.* **102** 308–24

Qu H, Zeng F, Tang Y, Shi B, Wang Z, Chen X and Wang J 2024 The clinical effects of brain-computer interface with robot on upper-limb function for post-stroke rehabilitation: a meta-analysis and systematic review *Disabil. Rehabil. Assist. Technol.* **19** 30–41

Rosenfeld J V 2024 Neurosurgery and the brain-computer interface *Adv. Exp. Med. Biol.* **1462** 513–27

Saway B F, Palmer C, Hughes C, Triano M, Suresh R E, Gilmore J, George M, Kautz S A and Rowland N C 2024 The evolution of neuromodulation for chronic stroke: from neuro-plasticity mechanisms to brain-computer interfaces *Neurotherapeutics* **21** e00337

Serrano-Amenos C, Hu F, Wang P T, Heydari P, Do A H and Nenadic Z 2024 Simulation-informed power budget estimate of a fully-implantable brain-computer interface *Ann. Biomed. Eng.* **52** 2269–81

Tzimourta K D 2025 Human-centered design and development in digital health: approaches, challenges, and emerging trends *Cureus* **17** e85897

Waisberg E, Ong J and Lee A G 2024 Ethical considerations of neuralink and brain-computer interfaces *Ann. Biomed. Eng.* **52** 1937–9

Zhu H Y, Hieu N Q, Hoang D T, Nguyen D N and Lin C-T 2021 A human-centric metaverse enabled by brain-computer interface: a survey *IEEE Commun. Surv. Tutor* **26** 2120–45

IOP Publishing

Nanomedical Brain/Cloud Interface
Explorations and implications
Frank J Boehm

Chapter 5

Impact of a brain/cloud interface on humanity— gateway to the ten-billion-synapse world mind

Melanie Swan

> The only true voyage of discovery is not to visit strange lands but to possess other eyes.
> —Proust 1929 *In Search of Lost Time* V 349

Contributing to the concluding portion of this book on nanomedical brain/cloud interface (B/CI) technologies, the current chapter considers the greater impacts of these advances on humanity, and in particular, the potential use of a B/CI as the gateway to a ten-billion-synapse world mind. A 'ten-billion-synapse world mind' is not meant literally (as the number of synapses in a fully fledged individual or group cloudmind would be much higher), but as a general concept to denote the collaborative activity of a B/CI-based cloudmind (minds safely connected to the internet cloud via B/CIs for interactive purposes). For the general realization of the B/CI and especially for B/CI cloudminds, a robust hardware and software solution is required. This work proposes quantum computing as the hardware platform for the B/CI, together with a holographic control theory (based on the anti-de Sitter/conformal field theory (AdS/CFT) correspondence) as the lever for macroscale control of the quantum computing cloud environment (the AdS/CFT correspondence is a universal control theory that orchestrates macroscale-quantum domains), and a bioblockchain neuroeconomy as the operating software of the in-brain B/CI neuralnanorobot network. Overcoming hindrances to large-scale group collaboration is addressed to elaborate peak-performance cloudminds. The proximate objective of the B/CI is to map, monitor, cure, and enhance neural activity. At the more abstract level of everyday reality, the purpose of the B/CI is to facilitate human productivity, well-being, and enjoyment. The stakes are that B/CI cloudminds would make significant progress toward the achievement of a Kardashev-plus society that is able to marshal all tangible and intangible resources through mental and physical means.

doi:10.1088/978-0-7503-2144-0ch5   5-1

5.1 Brain/cloud interface cloudmind

5.1.1 Brain–computer interface technologies

The aim of this chapter is to provide a detailed vision of how the cerebral cortex of the human brain might be safely, securely, and seamlessly integrated with the cloud, which would be manifested as a B/CI cloudmind for deployment in a collaborative context. There are two different classes of brain–computer interface (BCI) technologies (table 5.1). The first is the core BCI, which already exists and allows individuals to use electrical brain waves to control prosthetic limbs and computer cursors. The second is the B/CI (human brain/cloud interface), which is a future technology that interfaces the human brain with the internet cloud on an individual and group basis.

5.1.1.1 Brain–computer interface

The core technology is the BCI (brain–computer interface). A BCI or brain–computer interface is a direct communication pathway between a wired brain and an external device (Nicolas-Alonso and Gomez-Gil 2012). In this type of interface, electrical brain waves are used to direct external behavior via electroencephalography (EEG). One of the first demonstrations of a BCI, performed in 1988, was to control a robot (Bozinovski *et al* 1988). BCIs may include neuroprosthetics such as cochlear implants (220 000 of which had been implanted worldwide as of 2010 (NIH 2011)).

5.1.1.2 Human brain/cloud interface

The proposed technology is a B/CI (human brain/cloud interface), which would safely connect the human brain with the internet cloud (Martins *et al* 2019). Such a B/CI would be based on neuralnanorobotics (medical nanorobots specifically designed to operate in the brain).[1] Fleets of such neuralnanorobots would comprise the B/CI, having controlled connectivity between neural cells and external data storage and processing. One of the first objectives of the B/CI would be to monitor the brain's ∼86 billion neurons and ∼242 trillion synapses for health purposes.

5.1.1.3 Brain/cloud interface cloudmind

A cloudmind would exist as a construct wherein one or more minds are connected to the internet cloud (Swan 2016, 2019). A cloudmind might comprise an individual

Table 5.1. BCI (B/CI) technology platforms and functionality.

BCI (B/CI) technologies	Functionality
Core BCI	Prosthetic limb and cursor control
Cloudmind B/CI (individual and group)	Productivity, well-being, and enjoyment

---

[1] Brain-based nanorobots are referred to as 'neuronanorobots' or 'neuralnanorobots' (Martins *et al* 2016, 2019)

mind operating on the internet or multiple human and machine minds participating in an interactive collaboration. 'Mind' generally denotes an entity with processing capability (not necessarily a biological mind that is conscious). An individual or group cloudmind might pursue a variety of activities related to productivity, well-being, and enjoyment. Minds would be (hypothetically for now) interfaced with the internet cloud through a B/CI (a brain-resident network of neuralnanorobots). By linking brains to the internet, B/CIs could allow individuals to be more highly connectable not only via communications networks but also to other minds, which could enable new kinds of learning and interactions. The term 'cloudmind' might also be used to refer to industrial robotics coordination networks of cloud-connected smart machines (Keenan 2017) or 'brainets,' other forms of human or animal cloud-connected minds (Martins *et al* 2019).

5.1.2 Brain/cloud interface, medical nanorobots, and neuralnanorobots

The B/CI is composed of neuralnanorobots, a network of medical nanorobots designed to operate within the human brain. Medical nanorobots are envisaged as nanoscale molecular machines (1×10^{-9} m), which have been proposed to complement native cells to perform medically related tasks in the body. Several species of medical nanorobots have been conceptualized and articulated, such as respirocytes (artificial red blood cells), clottocytes (artificial platelets), and microbivores (artificial phagocytes) (Freitas 2000, 2005, 2012). The estimated dimensions of each medical nanorobot are from ~1–3 microns in diameter (1000–3000 nm in diameter), with even smaller components, on the order of 1–10 nm in diameter (Freitas 2012). Such medical nanorobots would patrol the body for health monitoring and intervention (Fahy and Wowk 2015).

Three species of neuralnanorobots have been proposed, which correspond to the different phases of neural signaling: axonal endoneurobots, synaptobots, and gliabots (Martins *et al* 2019). Axonal endoneurobots would align with the neuron's axonal transmission of the electrical action potential, whereas synaptobots would aid in signal transmission across the synaptic clefts between neurons, and gliabots would support the glial cells that facilitate neural signaling. For delivery to the brain, neuralnanorobots would need to traverse the blood–brain barrier, enter the brain parenchyma, ingress into individual human brain cells, and automatically position themselves at the axon initial segments of neurons (endoneurobots), in proximity to synapses (synaptobots), and within glial cells (gliabots) (Martins *et al* 2019). Full B/CI deployment would consist of a one-to-one copositioning of neuralnanorobots with human brain cells (Martins *et al* 2012).

5.1.2.1 Brain/cloud interface applications: map, monitor, cure, and enhance
Four levels of B/CI applications can be outlined to map, monitor, cure, and enhance neural activity (table 5.2). The first step in monitoring the brain necessitates the mapping of the connectome. Connectome mapping is essentially the creation of a wiring diagram of the entire structural and functional information of the brain at the

Table 5.2. B/CI applications: map, monitor, cure, and enhance neural activity.

B/CI function	Neural activity objectives and tasks
Map	Connectome mapping to create a wiring diagram of the brain.
Monitor	Direct monitoring of the brain's 86 billion neurons and 242 trillion synapses.
Cure	Acute and chronic disease response, restoring lost or damaged functionality.
Enhance	Enhancing neural activity related to learning, attention, and memory.

appropriate temporal and spatial resolutions (Martins *et al* 2016). A digital high-resolution connectome-based map of the brain suggests that an individual brain might be simulated, including for functional, health-related, and enhancement purposes. (An example of state-of-the-art connectome research is the 3D mapping of the mouse brain at single-cell resolution by the Allen Mouse Brain Atlas project (Wang *et al* 2020).)

Second, using the digital connectome, the next phase of monitoring could take place. This might include tracking and communicating information, initially issuing alerts and conducting daily health checks, and later backing up memories. The third phase is curative, facilitating diagnosis and curing the approximately 400 conditions that affect the human brain. The curative phase might entail restoring lost or damaged functionality (particularly that related to neurodegenerative diseases and senescence). For example, to combat stroke, neuralnanorobots might provide directed electrical stimuli to the brain to dissolve blood clots using ultrasound (Marosfoi *et al* 2015). The farther-reaching potential of the B/CI is indicated in the fourth phase, namely enhancement applications that could accentuate neural activity related to learning, attention, and memory.

The B/CI, and more generally, medical nanorobots, can be seen as a third onboard ecosystem. The human organism already consists of underlying human cells plus the microbiome (Yong 2016), whereas nanorobots would be an additional onboard ecosystem devoted to health monitoring, disease cure, and enhancement. The nanorobot ecosystem would be tasked with the provision of a protective buffer between the body's biological health and its environment.

5.1.3 Purpose of brain/cloud interfaces: the Kardashev-plus society

This section addresses the question of why B/CIs are needed. The proximate objective of B/CIs as a health-related tool to map, monitor, cure, and enhance biological capabilities is clear. However, the question arises as to what B/CIs mean from a more abstract level of human potentiality. It might be argued that the notion of a B/CI is not just novel but necessary to keep pace with the scale, velocity, and complexity of modernity. The world is evolving at an accelerated rate, and the capacity of the human brain to keep up has lagged. New strategies for understanding, learning, and coordination are essential. For example, B/CIs might enable direct neural transfer (transferring information directly into the brain) as a means of heightened learning (Martins *et al* 2019).

The overall aim of B/CIs may be seen as parallel to that of science and technology, whose general purpose is to contribute to the productivity, well-being, and enjoyment of society. Such a tripartite objective can likewise be the goal of B/CIs, with the rationale behind their implementation being to enable lives that are more meaningful, rewarding, and fulfilling. B/CIs are just one of many possible tools and technologies that might be deployed in support of the overall objective of the long-term survival and well-being of human society.

5.1.3.1 Kardashev civilizations

A large-scale vision for the progression of human society is Kardashev civilizations. Developed by Soviet astrophysicist Kardashev in 1964, this schema describes a society's ability to control energy-related resources and defines three levels of advancement (table 5.3). A Type I planetary society is able to utilize all of the energy from the sunlight that falls on the planet. A Type II stellar civilization is able to use all of the energy that the sun produces. A Type III galactic civilization is able to employ the energy of the entire galaxy. In terms of contemporary progress, one estimate posits that humanity has attained a Type 0.7 civilization and might advance to Type I within 100–200 years if energy consumption were to increase by 3% each year (Kaku 2018).

5.1.3.2 Kardashev-plus society

Kardashev's vision of the ability of civilizations to control energy resources is well known and can be expanded more generally to the notion of a Kardashev-plus society. A Kardashev-plus society is one that is able to marshal *all* resources, tangible and intangible, not only energy as a central resource for society's long-term success. To achieve a Kardashev-plus society, new models, techniques, and strategies such as B/CIs will be necessary. Tools that are adequate for carrying out a new tier

Table 5.3. The Kardashev scale: measuring the technological advancement of civilization by energy marshaling.

Civilization	Energy marshaling	Energy consumption	
Type I: Planetary civilization	Uses all the energy of the sun that falls on the planet	10^{16} W	$\approx 4 \times 10^{19}$ erg s^{-1} (4×10^{12} watts)^2
Type II: Stellar civilization	Uses all the energy that the sun produces	10^{26} W	$\approx 4 \times 10^{33}$ erg s^{-1} (4×10^{26} watts); the luminosity of the sun
Type III: Galactic civilization	Uses the energy of the entire galaxy	10^{36} W	$\approx 4 \times 10^{44}$ erg s^{-1} (4×10^{37} watts); the luminosity of the Milky Way galaxy

---

[2] The erg (Greek *ergon*: work, task) is a unit of energy equal to 10^{-7} joules in the centimeter-gram-second system of units. Erg/sec is a unit of energy or work per second.

of very large-scale projects beyond today's megaprojects[3] are needed. The B/CI might be needed as a workhorse tool for the coordination of both mental and physical resources at planetary and extraplanetary scales.

5.1.3.3 Theoretical model development

The thesis of this work is that B/CIs constitute a next-generation technology for realizing a Kardashev-plus society for two reasons. First, B/CIs are necessary merely to keep pace with modern reality. Second, B/CIs enable new strata of capabilities in coordinated neural activities for controlling a much larger set of resources for societal advances.

A theoretical model of B/CI technologies as a causal approach to achieving a Kardashev-plus society is outlined in figure 5.1. The research question that frames this analysis is as follows: how do B/CI technologies (including individual and group cloudminds) impact our capacity to attain a Kardashev-plus society? The desired outcome (dependent variable) is a Kardashev-plus society, with the premise being that B/CI technologies (the independent variable) have a causal influence on the objective. Within B/CI technologies, two main factors (moderating variables) may influence the ability to produce a Kardashev-plus society, namely hardware and software. Hence, this work proposes a twofold hardware and software approach to B/CI technologies as the means of engendering progress toward the attainment of a Kardashev-plus society.

This chapter unfolds in five parts. First, it provides an overview of the B/CI project as the instantiation of neural signaling with neuralnanorobots. Second, quantum computing is proposed as the hardware platform for the B/CI. Third, a holographic control theory (based on AdS/CFT correspondence) is articulated to interface the quantum computing cloud with the macroscale reality of interacting with the B/CI. Fourth, a bioblockchain neuroeconomy is suggested as the operating software for the autonomous control of the in-brain B/CI neuralnanorobot network. Fifth, the coordinated operation of peak-performance cloudminds in group

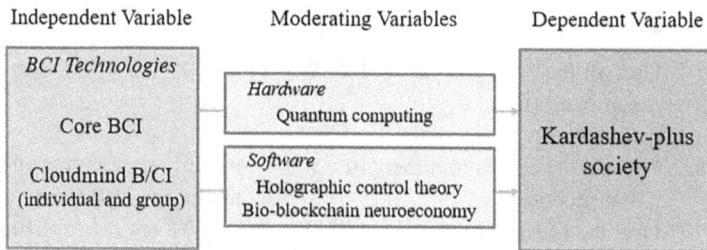

Figure 5.1. Theoretical model for attaining a Kardashev-plus society.

---

[3] Megaprojects are large-scale, complex ventures characterized by a large investment commitment, complexity, and long-lasting impacts on millions of people (Brookes & Locatelli 2015). Examples include bridges, airports, oil and gas extraction projects, hydroelectric facilities, nuclear power plants, and genome sequencing (Flyvbjerg 2017).

collaboration is addressed, wherein risks and limitations are considered. The appendices collect some technical details about neuralnanorobot size and B/CI implementation.

5.2 The brain/cloud interface project: neural signaling and neuralnanorobot instantiation

5.2.1 Summary of neural signaling

The human brain is comprised of an estimated 86 billion neurons, 242 trillion synapses (each neuron having ∼2800 synapses), and 85 billion glial cells (Martins *et al* 2019). A neuron is an electrically excitable cell that communicates with other cells by sending a signal called an action potential across specialized connections called synapses. Each neuron is comprised of a cell body (soma), a long thin axon insulated by a myelin sheath for outbound signaling, and multiple dendrites for receiving inbound signals. Glial cells are non-neuronal cells that insulate neurons from each other, facilitate signaling, and supply nutrients.

Neurons have two main processes: sending and receiving signals. To send a signal, an axon transmits information from the neuron to neighboring neurons. To receive a signal, a neuron's dendrites receive information sent by the axons of other neurons. The signaling activity of neurons is both electrical and chemical. The axons of neurons produce and transmit electrical pulses called action potentials, which travel along the axon like a wave. The action potential is a short electrical pulse that is 0.1 V in amplitude and lasts for one millisecond (Nicholls *et al* 2012). The action potential is sent along the axon to the axon terminals in the synaptic nerve endings, from which the axon contacts the dendrites of other neurons. Synapses (from the Greek word for conjunction) consist of a presynaptic terminal on the outbound neuron, a postsynaptic terminal on the dendrites of the receiving neuron, and a 20 nm gap between them (the synaptic cleft).

The electrical current responsible for the propagation of the action potential along the axons cannot bridge the synaptic cleft; thus, transmission across the gap between one neuron's axon and another's dendrites is accomplished by chemical messengers called neurotransmitters. Various chemical neurotransmitters are stored in vesicles (spherical bags) in the synaptic terminal at the nerve ending to be available for release across the synaptic junction.

At the presynaptic terminal (a bulbous area at the end of the neuron), the arrival of an electrical action potential causes voltage-gated calcium channels in the terminal wall to open and release calcium into the terminal bulb. The calcium triggers synaptic vesicles located in the terminal to release their neurotransmitter contents into the synaptic cleft. In less than a millisecond, the neurotransmitter diffuses across the gap and activates receptors in the membrane of the postsynaptic terminal (dendrites) in the receiving neuron. Glial cells are present around the synaptic cleft to facilitate signaling and to clean up, for example, by recycling neurotransmitters from the synaptic cleft back into synaptic vesicles (Shepherd 1974).

Although each neuron has only one axon (which ends in multiple axon terminals for sending signals), each neuron has multiple dendrites for receiving signals. On average, there may be ~2800 synaptic connections to other neurons (Martins *et al* 2019). Other estimates are higher, suggesting that each neuron may have an average of 7000 synaptic connections to other neurons (Finger 1994). Specialized Purkinje cells in the cerebellum have over 1000 dendritic branches, each with thousands of synaptic connections to other neurons. Synapses can be either excitatory or inhibitory, to reinforce or dampen the signal that emanates from the axon.

5.2.1.1 Size and scale (Avogadro's number)

Comprehending the B/CI requires thinking in various numeric scales (see the appendix for details and references). Some of the units employed are the micron (μm, a millionth of a meter, 1×10^{-6} m) and the nanometer (a billionth of a meter, 1×10^{-9} m). One micron is 1000 nm. As a heuristic, the diameter of a human hair is ~100 microns or 100 000 nanometers, and the basic nanorobot is 1000 nm in diameter. A red blood cell has a diameter of 7000 nm. Neurons are larger, having a 10 000–25 000 nm cell body, and glial cells are about the same size as neurons, at 15 000–30 000 nm. The synaptic terminal is quite small, perhaps only 100–1000 nm^3, and the synaptic cleft between neurons is even smaller at 20 nm. Glial cells are separated from each other by only 2 nm. The nanometer scale (1×10^{-9} m) is the scale of atoms and quantum mechanics. The atomic scale is employed when describing the (yet-to-be-realized) atomically precise molecular manufacturing of nanorobots and quantum computing that manipulates quantum entities (atoms, ions, and photons).

Estimates of the relative size of the different populations of neural cells are presented in table 5.4. Of the approximately 86 billion neurons in the brain, the 16 billion neurons that reside within the cerebral cortex may be of primary interest for a B/CI, as they are associated with higher-order functionality such as planning, reasoning, and vision.

Although 86 billion neurons, 85 billion glial cells, and 242 trillion synapses might sound like immense numbers, they may be more manageable in contrast to the overall context of the human body and the routine processing capabilities of data analytics. A 'really large number' in biology and chemistry is Avogadro's number, a trillion trillion (more specifically (6×10^{23}) or (0.6 of a trillion $\times$ a trillion)), which is used to refer to molecular volumes. Some of the initial tasks allocated to B/CI neuralnanorobots would be to catalog, assign specific spatial coordinates to, and track the billions and trillions of neurons, glial cells, and synapses. The next steps may involve modeling their activities, which will be more on the order of Avogadro's number. Quantum computation is envisioned to operate at the scale of Avogadro's number.

A quantum computer with 79 entangled qubits (current systems possess 20 qubits) would have an Avogadro number of states (with quantum entanglement, n qubits represent 2^n different states with which the same calculation can be performed simultaneously). This capacity is specified for large-scale computation environments (similar to that required by the B/CI), such as particle accelerators. CERN, for

Table 5.4. Sizes of neural cell populations in the human brain.

Entity	Size estimate		References
Neurons	86×109	86 000 000 000	Martins *et al* (2019)
Cerebellum (80%)	69×109	69 030 000 000	Azevedo *et al* (2009)
Cerebral cortex (19%)	16×109	16 340 000 000	Azevedo *et al* (2009)
Glial cells	85×109	85 000 000 000	von Bartheld *et al* (2016)
Synapses	2×1014	242 000 000 000 000	Martins *et al* (2019)
Avogadro's number	6×1023	600 000 000 000 000 000 000 000	Nelson (2008)

The value 242 trillion and 0.6 trillion × 1 trillion appear associated with Synapses and Avogadro's number respectively.

example, anticipates computation at the scale of Avogadro's number in the next upgrade of the Large Hadron Collider (LHC), the High-Luminosity Large Hadron Collider (HL-LHC), which is expected to begin operations in 2026, with an estimated required computational capacity 50–100 times greater than the capacity that currently exists (Carminati 2018).

5.2.2 Neural cells and neuralnanorobot complements

Three neuralnanorobot species have been proposed, which respectively correspond to the different phases of neural signaling, the cellular processes of neurons, and the subcellular processing of synapses. A one-to-one relationship is envisioned for the pairing of each neuralnanorobot with the neural cell to which it corresponds (table 5.5). Axonal endoneurobots relate to the axon and its function of sending action potentials as outbound signals. This class of neuralnanorobot may be located in the cell body (soma) at the beginning of the axon, which has an area of 10 000–25 000 nm; thus, the nanorobot can be relatively large (the standard 1000 nm is not a problem) and have a variety of functions.

Synaptobots correspond to synapses, which are the specialized interfaces between the nerve endings of neurons, as described above. Synaptobots may be embedded in the presynaptic terminals at the ends of axons to monitor outbound signals, the postsynaptic terminals at the ends of dendrites to receive inbound signals, or in intimate proximity to synaptic clefts. Therefore, synaptobots would need to be much smaller than axonal endoneurobots. The volumes of synaptic terminals range from 100 to 1000 nm^3, whereas synaptic cleft gaps are $\sim$20 nm; thus, the estimated dimensions of synaptobots would be on the order of $\sim$5 to 300 nm, with limited functionality. These nanodevices would be delivered to synapses in a phased implementation, for example, in a series of 117 shipments (Martins et al 2019). The glial cells that surround neurons are equivalent in size to neurons ($\sim$15 000–30 000 nm), which suggests that gliabots might also be fairly large ($\sim$1000 nm) and support a variety of functions.

Table 5.5. Neural cells and neuralnanorobot complement.

Neural cells	Function	Neuralnanorobot	Number of nanorobots	
Neurons				
Axon beginning (soma (cell body))	Sends signals	Axonal endoneurobot	1/neuron	86 billion
Axon ending (presynaptic terminal)	Sends signals	Synaptobot	2800/neuron	242 trillion (86 billion · 2800)
Dendrite (postsynaptic terminal)	Receives signals	Synaptobot		
Glial cells	Facilitates signals	Gliabot	1/neuron	85 billion

5.2.2.1 Neural signaling: electrical and chemical processes

The sophistication of neurons in converting electrical and chemical signals is notable. Similar to global telecommunications networks that convert between electrical and optical signals for efficient transmission, neurons convert signals between the electrical and chemical domains for efficient transfer (table 5.6). A neuron receives signals via dendrites and soma (for example, from visual or motor stimuli) and sends them down the axon as electrical action potentials. The action potential is converted from an electrical signal to a chemical signal in the presynaptic terminal and crosses the synaptic cleft as a chemical signal, which is processed within the postsynaptic density of the dendrites. The general form of neural signaling is through chemical synapses as described; however, electrical synapses are used exclusively for high-stakes, rapid-signaling applications, such as in the heart and for escape reflexes.

Although neural signaling is a complex electrical and chemical process, initial B/CI iterations may be realized on the basis of electrical signaling alone, without the involvement of chemically based neurotransmitters (Martins *et al* 2019). This implies the deployment of only the axonal endoneurobot. The implementation of a mature B/CI would also encompass the chemical operations of neurotransmitters.

Biomimetic principles could be employed to convert analog chemical signals to digital electrical signals for more efficient transfer. For example, optogenetic therapies bypass expensive rod-and-cone retinal processing by transmitting already converted electrical signals (Williams 2017). Neuralnanorobots, their neural cell complements, and electrical and chemical signaling functions are outlined in table 5.7.

The axonal endoneurobot will register the electrical activity of an axon that sends an action potential. The synaptobot in the presynaptic terminal might register electrical activity based on the calcium discharged into the terminal as the arrival of an action potential causes the voltage-gated calcium channels in the terminal wall to open. Synaptobots positioned at the synaptic cleft or postsynaptic terminal would need to register chemical neurotransmitter activity. Glial cells have resting potential but operate primarily with neurotransmitters; thus, the gliabot would also need to register chemical neurotransmitter activity.

Table 5.6. Neuronal signaling steps and signal types (E = Electrical, C = Chemical).

	Neural signaling step	Signal type (electrical or chemical)	
1	Axon produces action potential	Electrical pulse	E
2	Axon ending: action potential received at presynaptic terminal	Electrical-to-chemical neurotransmitter conversion	E-to-C
3	Signal transmitted across synaptic cleft (20 nm, 1 millisecond)	Neurotransmitters released into synaptic cleft	C
4	Dendrite: signal received at postsynaptic terminal	Neurotransmitter reception	C

Table 5.7. Neuralnanorobots and the required electrical and chemical transmission functionalities.

Neuralnanorobot	Neural complement	Electrical	Chemical	Function
Axonal endoneurobot	Axon	Action potential	N/A	Send action potential
Synaptobot				
Axon ending: presynaptic terminal	Presynaptic terminal	Action potential	Neurotransmitter	Electrical-to-neurotransmitter conversion
Synaptic cleft	Synaptic cleft	N/A	Neurotransmitter	Transmits neurotransmitters
Dendrite: postsynaptic terminal	Postsynaptic terminal	N/A	Neurotransmitter	Neurotransmitter-to-electrical conversion
Gliabot	Glial cell	N/A	Neurotransmitter	Facilitates neurotransmitter operations

5.2.3 Neurocurrencies

A *neurocurrency* is a resource that is employed to execute a neural function, either via a neural cell or neuralnanorobot. Each neuralnanorobot may conduct a variety of autonomous or semiautonomous activities that require the monitoring and allocation of resources. Such activities could be instantiated as transactions with resource balances that can be exchanged in an economic system such as a blockchain. Each activity is a transaction in which resources may be impacted. Transactions could include a small service charge (1% of the transaction value is standard) to support the system cost of the B/CI network and to encourage cost-based resource use. Such bioeconomic concepts have been explored in science fiction, in the contemplation of the body's future smart cells, mechanocytes, that self-liquidate if not kept balanced (Stross 2013). Various neurocurrencies and their relevance to neuralnanorobots are listed in table 5.8.

The B/CI might be manifested as a multicurrency environment, which could be denominated in various neurocurrencies as the coordination mechanism in the blockchain economic system. Some of the different categories of neurocurrencies might include electricity, ions, neurotransmitters, and fuel. Neurocurrency categories are generic wrappers for the different classes of resources transmitted in the neural signaling process and other operations; for example, the neurotransmitter class is a generic wrapper for the various neurotransmitters that might be exchanged.

5.2.3.1 Electricity and ions

The primary neurocurrency is electricity, where electrical balance may be interpreted as voltage, polarization, action potential, and resting potential. A related neurocurrency that neural operations traffic is ions (atoms stripped of one electron).

Table 5.8. Neurocurrencies by neuralnanorobot species.

Neurocurrencies ($NC)		Neuralnanorobot species		
Category	Resource	Axonal endoneurobot	Synaptobot	Gliabot
Electricity	Voltage	X	X	
	Polarization	X		
	Action potential	X		
	Resting potential			X
Ions	Sodium (Na^+)	X		
	Potassium (K^+)	X		
	Calcium (Ca^{2+})		X	
	Chloride (Cl^-)		X	
Neurotransmitters (Nx)	Glutamate (excitatory)		X	X
	Gamma-aminobutyric acid (GABA) (inhibitory)		X	X
Fuel	Glucose (adenosine triphosphate, ATP)	X	X	X
	Oxygen (ATP)	X	X	X

The most important neural cell ion balances are related to sodium (Na^+), potassium (K^+), calcium (Ca^{2+}), and chloride (Cl^-). Ion balances may be registered as positively charged ions (cations) or negatively charged ions (anions). Technically, ions are a form of electrical neurocurrency, given their charge determinations, but are functionally distinct enough to comprise a standalone neurocurrency. Each neuron maintains a voltage gradient across its membrane due to metabolically driven differences in sodium, potassium, chloride, and calcium ions, which should be measured separately.

5.2.3.2 Neurotransmitters

The third important neurocurrency is the neurotransmitter, i.e. chemical messengers, of which there are over 200 in total. Glutamate and gamma-aminobutyric acid (GABA) are the two most common neurotransmitters in the brain (accounting for 90% of all activity) and correspond to excitatory and inhibitory action, respectively. The third most notable neurotransmitter in neural signaling is acetylcholine (which increases the probability of presynaptic neurotransmitter release).

One canonical means of classifying neurotransmitters is by distinguishing large amino acids from small molecules (table 5.9). There are only a few large amino acid neurotransmitters, for example, glutamate and aspartate (excitatory), GABA and glycine (inhibitory). There are many small-molecule neurotransmitters including acetylcholine, dopamine, norepinephrine, histamine, serotonin, and epinephrine.

Table 5.9. Neurotransmitter classes and neuralnanorobot correspondence. (Table created by the author.)

Neurotransmitter class		Stimulation		Neuralnanorobot	
Type	Neurotransmitter	Excitatory	Inhibitory	Synaptobot	Gliabot
Large amino acids	Glutamate	X		X	X
	Aspartate	X		X	X
	GABA		X	X	X
	Glycine		X	X	X
Small molecule	Acetylcholine	X	X		X
(monoamine)	Serotonin	X	X		X
	Noradrenaline		X		X
	Dopamine	X			X

Small-molecule neurotransmitters may have the capacity for both excitatory and inhibitory actions, or just one. The well-known neurotransmitters acetylcholine and serotonin can have either excitatory or inhibitory actions, since they are related to general alert systems. Others, such as noradrenaline, signal stress, while dopamine signals the reward system. The large amino acid neurotransmitters are implicated in synaptic activity and therefore also for synaptobots, whereas the small-molecule neurotransmitters are more suitable for gliabot activity.

Another canonical strategy for classifying neurotransmitters is to determine whether they are ionotropic or metabotropic. Both operate by binding to trans-membrane-based receptors on the receiving neuron. Ionotropic neurotransmitters induce an ion channel to open in the receiving membrane, whereas the less powerful metabotropic neurotransmitters trigger a signaling cascade within the receiving cells (by coupling to G-proteins). Large amino acid and small-molecule neurotransmitters may be either ionotropic or metabotropic. For the implementation of a B/CI, ionotropic neurotransmitters are more functionally important and straightforward to model initially, as they lack the complex G-protein-coupled signaling cascade within the receiving dendrite.

5.2.3.3 Fuel
Fourth, but not least, a fuel balance is necessary for neural cells to conduct activities, namely by generating adenosine triphosphate (ATP) from oxygen and glucose delivered by the bloodstream. ATP is the energy currency of all cells. In neurons, ATP is required to drive the flow of charged ions that underlie electrical signaling activities. About two-thirds of a neuron's energy is used to produce sodium/potassium ATPase, an enzyme that recharges the ionic gradients of sodium and potassium after an action potential has occurred (Morris and Fillenz 2003). Likewise, synapses operate as a function of the ATP-based energy of mitochondria that circulate within the synaptic terminal.

High cytosolic calcium in the axon terminal triggers mitochondrial calcium uptake, which, in turn, activates mitochondrial energy metabolism to produce ATP that supports continuous neurotransmission. Synaptobots could track the fuel-based activities of the mitochondria. Fuel is also important for glial cells, as they mop up excess glutamate and other neurotransmitters via ATP-dependent pumps; thus, gliabots could track and predict neurotransmitter bursts.

A biodesign concern for neuralnanorobots is an autonomous energy source, meaning the ability of nanorobots to forage locally for the fuel necessary to sustain themselves and carry out operations. Estimates suggest that the standard nanorobot ($\sim$1000 nm) could produce several tens of picowatts of power from oxygen reaching its surface in the blood plasma (Hogg and Freitas 2009). This would provide enough power for the steady-state activities of the nanorobot. If equipped with pumps and tanks for onboard oxygen storage, the nanorobot might be able to collect enough oxygen to support burst power demands two to three orders of magnitude larger.

5.3 Brain/cloud interface hardware: quantum computing for the brain (quantum brain/cloud interface)

Living things are made of atoms according to the laws of physics, and the laws of physics present no barrier to reducing the size of computers until bits are the size of atoms and quantum behavior holds sway.

(Feynman 1985, Feynman *et al* 2005).

5.3.1 Communication and connectivity platforms

An overarching theme in the decades-old development of modern science and technology is the evolution of platforms used for connectivity and communication. There continue to be new and improved technologies for connecting humans to the internet cloud and to each other. Three key connectivity platforms might be noted: the personal computer, the smartphone, and potentially, in the future, the B/CI (brain/cloud interface), as shown in table 5.10. Equally important to the realization of the connectivity platform is the back-end network that supports it.

Highlighting the difficulty of forecasting the impacts of technology, IBM CEO Thomas J. Watson famously observed in 1943 that 'I think there is a world market for maybe five computers' (Strohmeyer 2008). Indeed, the market for expensive

Table 5.10. Evolution in connectivity platforms.

Eras of connectivity	1970–2020	2007–2025	2030–2050
Connectivity platform	Personal computer	Smartphone	Brain/cloud interface
Back-end network	Computer data networks	Global fiber-optic communications networks	Quantum computing networks

room-sized computers may have been small, but the trend that led to the personal computer and the internet has had an immense impact. It is clear that the overall intensity of human connectivity and communication has been increasing over time and that the platform of choice continues to miniaturize into forms that have greater portability and onboard convenience for humans.

Estimates in Cisco's Annual Internet Report suggested that by 2023, 67% of the global population would have internet connectivity (up from 51% in 2018), whereas 70% would possess mobile phones (up from 66% in 2018) (Cisco 2020). Internet-connected devices were forecast to amount to more than three times the global population (3.6 networked devices per capita, up from 2.4 in 2018). The key point is the remarkable jump in internet penetration in only five years (from 51% globally in 2018 to two-thirds in 2023) as smartphones proliferate. This suggests that future advances in connectivity platforms (such as the B/CI) might experience similar accelerated adoption patterns, further expanding the overall percentage of the global population that is connected to the internet.

It is clear that the prospect of a B/CI will necessitate the development of leading-edge technologies and may perhaps even inspire and drive them. The brain is one of the most complex systems known; thus, the latest research for its understanding and manipulation is required. Existing projects such as whole-brain emulation use technologies such as supercomputing to enable massively parallelizable cortical column simulations. However, a vastly more scalable platform than supercomputing (e.g. quantum computing) will be necessary for the next-level elucidation of the brain's inherent complexity (Harris and Kendon 2010).

Since the chemical conversion processes that occur within the brain are orchestrated by quantum mechanical principles, an analogous computing environment for representing these processes is required. The advantage of quantum computers is that they can natively store and process data associated with simulated quantum systems, thus providing tremendous scalability (Olson *et al* 2016). To perform quantum simulations, the Hilbert space (three or higher-dimensional space) of the underlying system is mapped directly onto the Hilbert space of the (logical) qubits in the quantum computer (Kendon *et al* 2010).

Quantum computing was initially indicated for the realization of B/CIs due to the sheer technical scalability in processing offered by this computational substrate, which has congruency with the human brain (e.g. the quantum mechanical chemical exchanges through which the brain operates). Second, quantum computing addresses the security sensitivity of B/CIs via nature's slate of security features built into the quantum mechanical domain, such as the no-copying and no-measurement principles.

5.3.2 Quantum computation

Quantum computing could lead to whole new regimes of bioinspired engineering at the nanoscale.

(Harris and Kendon 2010)

Quantum computing (data processing at the quantum scale of atoms (1×10^{-9} m)) is an early-stage but rapidly advancing technology (Swan *et al* 2020), and the field is further along than might be thought. Commercial systems (on-site and cloud-based) are shipping from three vendors: IBM, Rigetti (controllable gate model super-conductors with 19 qubits), and D-Wave Systems (less controllable quantum annealing machines with 2048 qubits). Advances in superconducting materials have enabled the production of superconducting chips that do not need to be cooled with bulky cryogenic equipment. Quantum computing is implicated in eventually being able to break existing cryptographic standards (2048 bit RSA), which is expected to lead to a 'Y2K for crypto.'

A 2019 U.S. National Academies of Sciences report estimated that this is unlikely within ten years (Grumbling and Horowitz 2019); however, techniques are con-tinuously improving. The U.S. National Institute of Standards and Technology (NIST) is developing next-generation standards based on lattice cryptography (complex 3D arrangements of atoms), as opposed to the difficulty of factoring large numbers (currently used by RSA 2048); this represents a mathematical shift to group theory (lattices) from number theory (factoring).

Quantum computing delivers an improved capacity to manipulate 3D reality at the atomic scale, which could make B/CIs more of a realizable possibility in the near term. As a broad heuristic, quantum computing may allow a one-tier increase in the computational complexity schema (i.e. the computational resources required to compute a given problem). For example, in the canonical traveling salesperson problem, it may be possible to check twice as many cities in half the time using a quantum computer.

A problem that requires exponential time in classical systems (i.e. too long for practical results) may take polynomial time in quantum systems (a reasonable amount of time). The improved performance of quantum computing is due to the superposition, entanglement, and interference (SEI) properties of quantum objects. In particular, superposition facilitates an acceleration in processing, as all problem inputs can be calculated simultaneously. All permutations of zero and one in 3D space can be tested concurrently until they collapse into one final answer (zero or one) at the end of the computation.

5.3.2.1 Quantum photonic networks: superposition of superposition
The standard one-tier speed-up in the computational complexity of quantum computing is due to the massive parallelism of being able to process all problem inputs simultaneously (all permutations of zero and one in 3D space). Superposition essentially means 'try all possibilities simultaneously.' Optical quantum computing offers an additional tier of computational speed-up: as in quantum photonics, there can be a superposition of both problem inputs and processing gates ('a superposition of superposition') (Procopio *et al* 2015). The difference between standard quantum computing and optical quantum computing lies in their treatment of gate architecture.

Standard quantum architecture has a fixed gate order (a fixed order of logic gates through which computation proceeds). Photonic quantum architecture, however, can have a superimposed gate order (that tests all possible permutations of gate order during the computation). In quantum photonics, there is a superposition of both form and content (gates and inputs), whereas in standard quantum computing, there is only a superposition of content (inputs). The result is that optical quantum circuits may offer an exponential advantage over classical algorithms and a linear advantage over standard quantum algorithms.

Quantum photonics might deliver greater computational capabilities, as well as next-generation global telecommunications networks. Optical networking is at the center of global communications networks today, and quantum photonics is anticipated as an upgrade technology. Although there are many ways to generate qubits (quantum information bits) for quantum computing on standalone machines, for a larger architecture of networked machines, electrical signals must be converted to optical signals. Hence, there are two methods in development for quantum computing networks. One way is to create an all-optical platform from the onset with continuous qubit optical interfaces (Sapra *et al* 2020). The other method is to design a microwave superconducting platform (using the semiconductor template) that is subsequently interfaced to optical networks with electrical–optical interconnects.

Quantum photonics is further indicated for the construction of a quantum internet. The quantum internet is a proposed next-generation internet that would feature secure end-to-end communications as an improvement upon the lack of privacy and security of the current internet. The quantum internet would employ quantum switches and routers, quantum key distribution, quantum processors, and quantum memory. One possible roadmap for migrating to the quantum internet was set forth by Wehner *et al* (2018). If quantum photonic networks are to become the secure global platform of choice for computing and connectivity, entirely new classes of Kardashev-plus applications might become possible, such as the realization of the B/CI.

5.3.2.2 Quantum neuroscience and brain/cloud interface applications

Thus far, the demonstrated applications of quantum computing are in the areas of optimization and simulation. Optimization and simulation could have an important impact on the potential realization of B/CIs, whole-brain emulation, and computational neuroscience. In classical computing, state-of-the-art progress includes a proposed brain–machine interface platform with 3072 electrode-based channels (Musk 2019), and a partial brain simulation of 80 000 neurons and 0.3 billion synapses, as a sample representation of the 86 billion neurons and 242 trillion synapses in the brain (van Albada *et al* 2018).

Quantum computation has a great deal to offer to the complexities of neuroscience. For example, the 'test all permutations' functionality of superposition could enable a more tractable approach to brain-related calculations. In the standard compartmental model of neural signaling, a single neuron might have a thousand separate compartments whose behavior might be described by tens of thousands of differential equations. The differential equations represent the numerical integration of thousands of nonlinear time steps and summarize the activity of counting the voltage spikes (threshold signals) that emanate from the neuron. This type of modeling project becomes much more feasible with quantum computing. A deployment of this idea has been proposed for modeling the pyramidal cell (one of the most sophisticated neurons) in a dendritic tree model abstracted into a two-layer neural network (Poirazi et al 2003).

Neural signaling is often analyzed using the Hodgkin–Huxley[4] model, which describes the conduction of the electrical impulse through the axon. The Hodgkin–Huxley model has been studied in the quantum regime with a hardware model that incorporates the three axon ion channels (potassium, sodium, and chloride), together with a signal source and output, using a system of memristors, resistors, and capacitors (Gonzalez-Raya et al 2020). This work provides a blueprint for the construction of quantum neuron networks with quantum state inputs. This might facilitate the realization of a B/CI via quantum computing (a quantum B/CI) and, more generally, hardware-based neuromorphic quantum computing (e.g. large-scale systems that mimic the neurobiological architectures of the human nervous system in a computing environment (Monroe 2014)).

5.3.3 Nature's built-in quantum security features

The previous section discussed ways in which quantum computation offers new levels of capabilities, which may enable a variety of projects, including the realization of the B/CI. This section discusses security features that are specific to quantum computing (Swan 2020). Five specific security features inherent to quantum mechanical domains are as follows: the no-cloning theorem, the no-measurement principle, quantum error correction through particle entanglement, quantum statistics (provable randomness), and computational verification owing to the bounded-error quantum polynomial time (BQP) computational complexity class of quantum information (table 5.11).

First is the no-cloning theorem, which states that quantum information cannot be copied, meaning that there is fidelity and uniqueness in any information transfer. Second is the no-measurement principle, which states that quantum information cannot be viewed or measured without changing or damaging it. This means that eavesdropping is immediately detectable (and would practically trigger a resending of the quantum information with slightly different encoding). Third is the fact that entangled particles allow the realization of quantum error-correction schemes.

Fourth is quantum statistics in that distribution signatures could have only been quantum-generated. Certain quantum statistical distributions based on the SEI properties of quantum objects have specific wave motion that could only have been generated by quantum systems, thus conveying provable randomness.

Fifth is the computational complexity class of quantum information (BQP/ quantum statistical zero-knowledge (QSZK)). The computational complexity class of problems that can be solved with a quantum computer is known as BQP, which is

---

[4] The 1963 Nobel Prize in physiology or medicine was awarded for their description of the propagation of electric signals in squid axons.

Table 5.11. Natural security features built into quantum mechanical domains.

	Principle	Quantum security feature
1	No-cloning theorem	Cannot copy quantum information.
2	No-measurement principle	Cannot measure quantum information without damaging it (eavesdropping is immediately detectable).
3	Quantum error correction	Error correction via ancilla (larger state of entangled qubits).
4	Quantum statistics	Provable randomness: distributions could only be quantum-generated.
5	BQP/quantum statistical zero-knowledge (QSZK) computational complexity and computational verification	Quantum information domains compute quickly enough to perform their own computational verification (zero-knowledge proofs).

contained within the computational complexity class QSZK. 'Zero knowledge' refers to a mathematical soundness attribute in which it is not necessary to have any knowledge of an underlying process (i.e. zero knowledge), only the result. Quantum computers operate quickly enough to perform their own computational verification (zero-knowledge proofs) of the results as part of their operation. Together, these native security features of quantum information domains suggest that quantum computing provides an extremely secure processing environment.

5.4 Brain/cloud interface software: holographic control theory

The previous section elaborated on quantum computing as the hardware platform for the B/CI. The next sections discuss the control software and the operating software. For the control software, a holographic control theory (based on the AdS/CFT correspondence) is proposed as a control theory to bridge the macroscale and quantum domains (B/CI computations are carried out in a cloud-based quantum computing environment). B/CI neuralnanorobot network data is collected and computed in quantum mechanical form and abstracted for practical use by human administrators at the macroscale. For the operating software, a blockchain-based neuroeconomic system is proposed to coordinate the in-brain B/CI neuralnanorobot network.

5.4.1 Holographic correspondence (the anti-de Sitter/conformal field theory correspondence)

Considering software for the B/CI quantum computing cloud platform, a logical choice is a holographic control theory based on the AdS/CFT correspondence. Quantum reality (exemplified in this scenario by the brain and the B/CI) is a domain of quantum information processing; thus, an information-related control theory makes sense. The AdS/CFT correspondence is a known model for linking the quantum and macroscale domains. The AdS/CFT correspondence (also called holographic correspondence, gauge-gravity duality, and the bulk–boundary relation) is the claim that in any physical system, there is a correspondence between a volume of space and its boundary region such that the interior bulk region can be described by a boundary theory that reduces the number of dimensions by one. The AdS/CFT correspondence is a formalization of the holographic principle, which denotes the possibility of a 3D volume being reconstructed on a 2D surface.

5.4.1.1 Anti-de Sitter/conformal field theory correspondence

The central idea of the correspondence is that a messy bulk process running in a 3D volume can be written in a simplified manner as a surface theory on the boundary using one fewer dimensions. The AdS/CFT correspondence is analogous to a soup can in that the bulk is the 3D interior of the can, and the boundary is the 2D exterior label of the can. This is a useful visual representation of the difference between a 2D surface and a 3D bulk. However, more technically, AdS refers to anti-de Sitter space as opposed to regular de Sitter space.

Regular de Sitter space is the normal 3D space of lived reality, which can be described by Euclidean geometry. Anti-de Sitter space is a simplified model based on the hyperbolic geometry of a sphere and looks like the *Circle Limits* works of Escher that include pictures of fish and bats getting smaller and smaller as they extend toward the edge of a circle. Anti-de Sitter space is used in physics as a solvable model of de Sitter space. The benefit of the AdS/CFT correspondence is that a complicated bulk region can be resolved analytically and related to a boundary region using one fewer dimensions. In 'AdS/CFT,' CFT denotes conformal field theory, which means any basic (conforming) field theory.

5.4.1.2 Solving messy bulk systems and eliciting emergent structure

Applying the AdS/CFT correspondence renders what seems to be an intractable bulk volume in a manner that is solvable as a boundary theory in one fewer dimensions. For example, the overall universe could be considered as a bulk volume for which there might be a boundary theory that describes it using one fewer dimensions. Even without accessing the edge (i.e. getting outside the universe), the correspondence is a formal model that might be applied to derive information about the emergence of bulk structure, such as the distribution of matter and the development of space and time. Not only can a messy bulk volume be solved for practical use in one fewer dimensions as a surface theory, but an emergent structure might also be elicited to study the parameters of a system. For example, in many complex systems, the endpoint is known (e.g. gross domestic product, GDP) but not the patterns that arise along the way (e.g. merchant–consumer interactions) that define the result. The correspondence can be used as a technology to study complex systems.

Examples of complex systems with emergent structures include food web ecosystems, financial risk, social networks, and machine learning networks. In systemic financial risk, the process of contagion that leads to collapse is unknown. The ways in which influencers build traction in social networks are likewise targets of study. Machine learning is a black box in which it is unclear how the back-propagation and feed-forward network operations cycle to produce an optimal classification algorithm. Similarly, biological processes in the brain, and perhaps B/CI operations, are black-box systems that might be studied using correspondence as a complexity technology.

5.4.2 The black hole information paradox

The AdS/CFT correspondence was proposed by Maldacena (1999) as a formalization of the holographic correspondence developed by Susskind (1995) as a resolution to the black hole information paradox. The black hole information paradox asks how it is possible for information to exit a black hole. The conundrum arises from trying to understand quantum mechanics and general relativity together. The paradox is that on the one hand, information apparently cannot escape from black holes (per general relativity), but on the other hand, Hawking radiation does emanate from black holes. The only way that information (quantum information bits) can be contained in the Hawking radiation is if the information inside the black hole is copied, but the no-cloning theorem prevents quantum information from being copied (per quantum mechanics).

Susskind proposes the holographic principle (also called black hole complementarity) to resolve the black hole information paradox. The premise is that there are complementary views, both accurate, of the same physical phenomenon in the universe. This is not a stretch, as various time paradoxes (e.g. the grandfather paradox and the twin paradox) have been demonstrated in general relativity. They show that phenomena look different to different observers in the universe due to time dilation. In the case of the black hole information paradox, a far-off observer sees information smearing out on the event horizon surface of the black hole in 2D but not actually entering the black hole.

On the other hand, the nearby observer who is jumping into the black hole sees information in 3D, entering the 3D interior volume of the black hole. The complementary views suggest that there is no conflict, thereby resolving the paradox. The boundary (black hole event horizon) is described by a 2D surface theory, and the bulk (black hole interior volume) is described by a 3D volume theory. The cornerstone principle of the AdS/CFT correspondence is nicely demonstrated as a surface theory describing a messy bulk process in one fewer dimensions.

5.4.2.1 The anti-de Sitter/conformal field theory correspondence as gauge-gravity duality

Continuing to solve puzzles at the intersection of general relativity and quantum mechanics, Maldacena (2012) further instantiated the AdS/CFT correspondence as gauge-gravity duality. Gauge-gravity duality relates gravitational fields (general relativity) and gauge theory fields (quantum mechanics). Its underlying intuition is that both gravity and gauge theory are field-based phenomena at the scale of gauge theory (1×10^{-15} m) and therefore might be joined through the correspondence. While quantum mechanics (1×10^{-9} m) describes the behavior of atoms, gauge theory describes the behavior of subatomic particles (such as gluons and quarks) that comprise atoms.

At the scale of gauge theory, fields of flux may describe how force particles (bosons, such as gluons) hold subatomic matter particles (fermions, such as quarks) in place in atomic configurations. The AdS/CFT correspondence as gauge-gravity duality is an example of a foundational physics advance that may inform our understanding of the brain and the implementation of the B/CI. While brain activity is often considered at the scale of the quantum mechanical interactions of neurons $(1 \times 10^{-9}$ m), the subatomic particle scale of gauge theory $(1 \times 10^{-15}$ m) might illuminate deeper levels of the subcellular processing of synapses.

5.4.3 The anti-de Sitter/conformal field theory correspondence as a brain/cloud interface control theory

The AdS/CFT correspondence can be used as a control theory for a variety of systems, including those that span the macroscale and quantum domains. More generally, the correspondence is an example of a two-tier information system. Many two-tier systems have complex operating processes that must run in real time in the physical bulk to produce a final answer of interest in one fewer dimensions at the boundary (table 5.12). More formally, these systems can be described as one-way functions.

Two-tier information systems can be instantiated in bulk–boundary relationships using the AdS/CFT correspondence. Moreover, the AdS/CFT correspondence can be employed as a control theory to manage these processes, while surface theories can be used to control bulk processes. The insight of the correspondence is that any fleet (many-unit) domain can be orchestrated using the correspondence as a control theory. As a control theory, the correspondence allows many-particle domains to be managed via a temperature term. The AdS/CFT correspondence is a universal control theory that might likewise apply to taxis, spaceships, and B/CIs.

Table 5.12. Two-tier information systems with bulk–boundary relationships.

	Messy bulk process	Boundary output
1	Air particles moving in a room	Temperature
2	Merchant–consumer interactions	GDP
3	Quantum mechanical reality: particles jiggling	Macroscale reality: table
4	Information entering black hole interior in 3D	Information smeared out on the event horizon in 2D
5	Ancilla of larger entangled state	Error-corrected qubit
6	Hash function algorithm	Hash code output
7	Zero-knowledge proof	True/false value
8	Proof-of-work mining	Confirmed transaction block
9	Holographic annealing	Lowest energy state of a system
10	Protein folding process	Folded protein
11	Speculative exchange of oil (8–15x/barrel)	Consumed resource

B/CIs require a control theory that spans the macroscale of lived reality and the quantum domain of the brain–B/CI system. The AdS/CFT correspondence is an excellent candidate for such a B/CI control theory. B/CIs have fleets (many items) that need to be managed, which are the neuralnanorobots embedded *in vivo* within the brain that comprise the B/CI network. The B/CI domain is quantum mechanical, both in its physical reality in terms of neuralnanorobot operation and in its computational instantiation for control via the quantum computing cloud environment. The AdS/CFT correspondence easily spans the quantum scale and the macroscale and provides a surface-level theory for controlling messy processes in the quantum mechanical bulk.

5.4.4 The anti-de Sitter/conformal field theory correspondence as a control model for complex domains

The AdS/CFT correspondence is a universal control theory and particularly a control theory for complex domains. Complex adaptive systems are those that are nonlinear, emergent, open, unknowable at the outset, interdependent, and self-organizing. In the B/CI context, two aspects of complexity management are of interest: identifying novel structural emergence and providing an ambient control structure.

Identifying novel structural emergence is a key property of the AdS/CFT correspondence. The correspondence was initially developed to understand more about how the physical structure of the universe evolved—structures such as matter, geometry, space, and time. Structural emergence is the target of AdS/ML (machine learning) research, which aims to understand how the black box of deep learning systems settles upon certain algorithmic structures as being optimal. For the B/CI, one of the most important structural emergences is new ideas, as novelty in the form of ideation is a critical outcome for B/CI cloudminds, whether individual or collective.

Mechanisms for identifying new ideas as emergent structures via the holographic control theory could be essential for B/CI implementation. Positive emergent structures for B/CIs are new ideas. Negative emergent structures are also useful, as they signal problems developing in the B/CI as a complex adaptive system. Known aberrant behavioral patterns that emerge in automated fleets may likewise have an analog in B/CI networks. For example, in unmanned aerial vehicle (UAV) networks, thrashing, resource starvation, and phase change have arisen (Singh *et al* 2017), and in deep learning, vanishing gradients (disappearing problem horizons) are a known problem (LeCun *et al* 2015).

The provision of an ambient control structure is a second attribute of the correspondence-based control theory. The scope of operation of complex systems is, by definition, unknown at the outset, which consequently requires a flexible control model that evolves with the system. B/CIs need a control model; however, it is too early to precisely know the detailed requirements for such a control system. Therefore, it makes sense to select a flexible control model that may be able to handle any potential situations that arise.

Even the basics of B/CI operation are as yet unknown, such as the suite of onboard versus remote control features. Likewise, energy requirements and local resource availability are unknown. The 'temperature terms' that are most relevant for the B/CI are unclear. Furthermore, even once the relevant temperature terms have been identified, the corresponding 'thermostat' for their control will be required. Hence, the AdS/CFT correspondence is indicated as a flexible control model for the complexity of the B/CI and its operating environment.

5.4.5 Anti-de Sitter/conformal field theory correspondence studies

The AdS/CFT correspondence is a universal control theory that not only orchestrates macroscale-quantum domains but also complex systems. Given its management of messy bulk processes via a simplified boundary theory in one fewer dimensions, the correspondence is proving to be a foundational model with applications in many fields. Beyond cosmology and physics, some of the other disciplines employing the correspondence include materials science (strongly coupled domains such as superconductors, condensed matter, and plasma physics), machine learning, network theory, and blockchain distributed ledgers (table 5.13).

Table 5.13. AdS/CFT correspondence species and functionality.

AdS/CFT correspondence variation		Application functionality	References
AdS/CFT	AdS/CFT	Cosmology, particle physics	Maldacena (1999)
AdS/CMT	AdS/condensed matter theory	Strongly coupled systems: condensed matter, superconductors, plasma physics	Sergio and Pires (2014), Hartnoll *et al* (2018)
AdS/ML	AdS/machine learning	Elicit bulk structure as emergent neural network structure (weights, layer depth)	Hashimoto *et al* (2018)
AdS/DLT	AdS/distributed ledger technology	Holographic consensus, quantum smart routing, certified randomness	Kalinin and Berloff (2018)
AdS/B/CI	AdS/brain–cloud interface	Holographic backup, ad-hoc field assembly, and neurocurrency transfer	Swan *et al* (2020)
AdS/brain	AdS/brain	Holographic model of neural signaling	Swan *et al* (2022)

5.5 Brain/cloud interface operating software: bioblockchain neuroeconomy

This section elaborates a holographic control theory for the B/CI as a control lever that interfaces the macroscale of everyday physical reality with the quantum mechanical scale of the quantum computing platform. The holographic control theory coordinates between the B/CI neuralnanorobot network (with data collected and computed in quantum mechanical form) and its abstraction for practical use by human administrators at the macroscale. What remains is to characterize operating software for the B/CI neuralnanorobot network itself, i.e. to describe how the fleets of many-particle neuralnanorobots are to be orchestrated to perform their in-brain activities, and for this, a blockchain is proposed. A blockchain is an automated cryptographic system for value transfer and secure transaction logging (Swan 2015a). The holographic control theory can be implemented as a blockchain-based economic system. The correspondence is a control theory, i.e. a model for controlling a system, while the system to be controlled is the network of neuralnanorobots embedded in the human brain that comprise the B/CI.

5.5.1 Bioblockchain neuroeconomy

Economic concepts are the design principles of the B/CI network operating software. The B/CI neuralnanorobot network might utilize a bioblockchain (a blockchain deployed in a biological setting) as a transaction-logging system for the neuralnanorobots that have preprogrammed goal-directed behavior and carry out neurocurrency-based operations. A blockchain is anticipated for the B/CI operating software due to its technical feasibility and for security reasons. From a technical perspective, a modern smart network technology is needed that is transaction-heavy, network-based, and automated. The B/CI must orchestrate many-particle fleet units and their activities.

A blockchain is able to seamlessly register an arbitrarily large number of participants and likewise may be able to execute an arbitrarily large number of transactions. Thus, a bioblockchain economic system is a top choice for operating software for an automated transaction system. Blockchains support multicurrency environments, where in the case of the B/CI, neurocurrencies such as electricity, ions, neurotransmitters, and fuel are the basis for the execution of operations. Security and trust are created via the cryptographic features inherent in blockchains, such as real-time transaction confirmation. Finally, blockchains are a modular system that can easily scale for a B/CI cloudmind implementation, from individuals to groups. The same transaction-logging and security features are relevant in group cloudminds for intellectual property tracking, credit assignment, and privacy protection as for individual cloudminds.

5.5.2 Tech specs: brain/cloud interface neuralnanorobot network system requirements

The first function of the B/CI is to instantiate neural signaling in a computational system. To do so, the technical requirements are for a B/CI transaction system that

Table 5.14. Neuralnanorobot transaction volumes (estimated).

Neuralnanorobot class	Number of neuralnanorobots	Number of transactions (total)
Axonal endoneurobot	86 billion	1 s^{-1} or more $\times$ 86 billion
Synaptobot	86 billion $\times$ 2800 = 242 trillion	1 s^{-1} or more $\times$ 242 trillion
Gliabot	85 billion	1 s^{-1} or more $\times$ 85 billion

instantiates many-particle fleet units (neuralnanorobots) and their activities. This is on the order of 86 billion axonal endoneurobots, 85 billion gliabots, 242 trillion synaptobots, with average activity rates that may well exceed one transaction per second per unit (table 5.14). B/CI Phase I implementation is estimated to include only electrical signaling in the form of action potential impulse tracking performed by axonal endoneurobots. B/CI Phase II implementation would include the full inventory of addressable entities (also the 242 trillion synaptobots and 85 billion gliabots) engaged in electrical and chemical signaling processes that are both cellular and subcellular (synaptic processing). Initial B/CI deployment would likely focus on a specific brain region, function, or activity, with whole-brain engagement and functionality to follow.

The data processing requirements will entail enabling a B/CI to have controlled connectivity between the neural activities of neurons and synapses, together with external data storage and processing. B/CI neuralnanorobots would require an extremely fast wireless transmission capacity, on the order of 6×10^{16} bits s^{-1} (Martins *et al* 2019). The idea is to transmit synaptically processed and encoded human brain electrical information via auxiliary nanorobotic fiber optics (30 cm^3) with the capacity to handle up to 10^{18} bits s^{-1} and provide rapid data transfer to a cloud-based computing environment for real-time brain-state monitoring, data extraction, and exchange.

5.5.2.1 Brain/cloud interface neuralnanorobot network traffic

Table 5.15 outlines B/CI applications by traffic type, showing the relevant neuro-currency ledger units in which the transactions might be denominated, tracked, and exchanged. Existing applications offered by the core BCI include neuroprosthetics and computer cursor control. These activities are electrical transactions tracked by EEG signals, as the types of traffic being transmitted are denominated in microvolts. The application classes for the cloudmind B/CI encompass mapping the connectome, monitoring ongoing neural activity, preventing and curing diseases, and enhancing neural performance in learning, productivity, and experiential enjoyment.

As a communications platform, the B/CI network must accommodate diverse types of traffic. At a minimum, this might include internet protocol (IP) traffic, electrical signaling, and neurotransmitters. Just as the internet transfers various kinds of traffic using different software protocols (e.g. data, voice, and video traffic), so too the B/CI network will need to transfer different types of traffic related to its activities. Each traffic type could have its own software transfer protocol (i.e. operating instructions), neurocurrency, and measurement metrics.

Table 5.15. B/CI applications by traffic type and neurocurrency ledger units.

Application class	Application	Functionality	Traffic type	Ledger unit
Core BCI				
Current applications	Neuroprosthetics	Actuation	Electricity: EEG signal	Microvolts
	Cursor control	Actuation	Electricity: EEG signal	Microvolts
Cloudmind B/CI				
Map	Connectome	Functional mapping	Internet protocol (IP): 3D point cloud simultaneous localization and mapping (SLAM)	MB, spatial placement
Monitor	Data upload, backup	Security, privacy	IP: hypertext transfer protocol (HTTP) POST/GET	MB, service level agreements (SLAs)
Cure	Intervention delivery, preventive maintenance	Disease cure, rejuvenation	Electricity: ultrasound; pharmaceuticals, cellular therapies	Millivolts (mV), millimoles (mM), cells
Enhance	Transparent shadowing, direct neural transfer	Stimulaton, augmentation	IP: HTTP POST/GET	MB

Electricity is one neurocurrency, denominated in millivolts (mV), that allows the transfer of signals and the provisioning of field potentials. Another neurocurrency is neurotransmitters (e.g. serotonin, dopamine, and GABA), measured in nanomolar concentrations denominated in millimoles (mM). Internet traffic could mainly take the form of hypertext transfer protocol (HTTP) POST/GET requests for posting status and retrieving information, measured in MB of data transferred, along with the data transfer rate and service level agreements (SLAs). Connectome mapping has a specialized form of traffic to capture the 3D positioning data (e.g. triaxial spatial coordinates) of neural entities via simultaneous localization and mapping (SLAM) point clouds and could be measured in MB of 3D data. The main traffic types and neurocurrencies applicable to the neural signaling support operations of neuralnanorobots are listed in table 5.16. Different neurocurrencies may be employed to transfer electrical and chemical traffic within the B/CI network.

5.5.2.2 Brain/cloud interface neuralnanorobot network communication

Neuralnanorobots may communicate with the cloud, each other, and directly with neural cells (table 5.17). Interfacing with the cloud would be via IP traffic. Communication with other neuralnanorobots in the B/CI network would take place via IP traffic and neurocurrency balances. Direct interactions with neural cells would use native neurocurrencies (electricity, ions, neurotransmitters, and fuel). For communications between nanorobots, different models of biophysics-based chemical signaling have been proposed. For example, one nanorobot might guide other nanorobots to malignant tissues by issuing a higher intensity or gradient of E-cadherin as a chemical signal (Cavalcanti *et al* 2006). For direct communications

Table 5.16. Neuralnanorobots, biocurrencies, traffic types, and neurocurrencies.

Neuralnanorobot	Traffic type	Neurocurrency	Ledger unit
Axonal endoneurobot	Electricity	Electricity, ions	Millivolts (mV)
Synaptobot	Neurotransmitter	Neurotransmitter	Millimoles (mM)
Gliabot	Neurotransmitter	Neurotransmitter	Millimoles (mM)

Table 5.17. Neuralnanorobot communications: cloud, B/CI network, and neural cells.

Neuroanorobot communication	Traffic type	Task activity
1 To the cloud (two-way)	IP (internet protocol)	HTTP POST/GET
2 To other neuralnanorobots	IP & neurocurrency	Messaging, resource balancing, group coordination
3 To neural cells	Neurocurrency (electrical and chemical)	Polarization, voltage-gating; Neurotransmitter delivery

with neural cells, neuralnanorobots might use a combination of electrical signaling (to manage polarization, channels, and potential) and chemical signaling (to mediate neurotransmitter delivery).

5.5.2.3 Bioblockchain neuroeconomy implementation

To implement a bioblockchain neuroeconomy, the first step is to assign each entity a unique identification number (address) so that all activities can be logged by the system. An inventory of fleet units is thereby created. The next step is to assign operating goals, allowed functions, and resource balances to classes of units. Just like their real-life cell counterparts, neuralnanorobots could begin with basic neurocurrency balances (for electricity, ions, neurotransmitters, and fuel). Third, security, backup, and lifecycle management parameters could be assigned to groups of units. Finally, smart contracts (automatically executing programmatic instructions) could be instantiated for autonomous neuralnanorobot control, including the safety measure of an OFF switch to be used in the case of gross system failure.

Three specific phases of neuralnanorobot implementation can be articulated, corresponding to the instantiation, operation, and exception reporting of the B/CI network (table 5.18). To begin, entities are created, resource balances are assigned, and goals are applied. In the operating phase, neuralnanorobots perform tasks, their behavior is logged and monitored, their lifecycles are managed in terms of instantiation and retirement, and they ingress into and egress from their assigned locations within the brain. In the exception reporting phase, security procedures are followed and risk management is performed. Anomaly detection techniques from complexity science can be applied to identify unexpected behaviors that may develop in B/CI neuralnanorobot fleets.

5.5.2.4 Neural lightning network for neurocurrency replenishment

Biomimetic principles might be employed in the design of a resource recovery mechanism similar to that used by glial cells. The glial cell neurotransmitter recycling operation is analogous to payment channel resource rebalancing in blockchains. In blockchain economic networks, overlays such as the Lightning Network allow parties to precontract to automatically rebalance accounts (analogous to airline refueling contracts or replenishing a checking account from a linked savings account if the balance dips below a certain level) (Poon and Dryja 2016).

The B/CI neuronanorobt network could likewise run a payment channel system. Neuralnanorobots could contract with each other ahead of time as a feature of the

Table 5.18. Operating phases of the bioblockchain neuroeconomy.

Instantiation	Operation	Exception reporting
• Create entities	• Run	• Security
• Allocate resource balance(s)	• Log and monitor behavior	• Risk management
• Assign goals	• Lifecycle management	• Anomaly detection

blockchain-based smart contracts that orchestrate their normal operations. The contracts could instantiate dynamic resource rebalancing, meaning the agents (neuralnanorobots) would not need to constantly directly transact with each other in real time but could rather participate in an automated resource rebalancing scheme (still executed as unitary transactions, but carried out automatically).

Biomimetic principles might also be used to harness the group coordination feature built into neural signaling. The B/CI network could similarly encourage and reward multiagent behavior. For example, serotonin balances could be distributed to neuralnanorobots that link their activity to a group goal, such as improving the synaptic release of serotonin in signaling, with the macroscale result of reducing depression. The practical benefit could be to reduce the side effects of prescription drugs using the more granular native activation of neurotransmitters.

5.6 Peak-performance brain/cloud interface cloudminds

The previous sections have outlined the hardware and software platforms for the B/CI cloudmind. This section considers the challenges of implementing B/CI cloudminds in a group setting. Beyond the individual B/CI cloudmind (one mind connected to the internet cloud), there are issues specific to the implementation of group B/CI cloudminds (multiple minds safely and securely connected to the internet cloud via B/CI for collaborative activity, according to the concept of the 'the ten-billion-synapse world mind'). For the realization and peak performance of B/CI cloudminds, effort is required along two implementation trajectories: one to instantiate well-formed groups and the other to overcome hindrances specific to large-scale group collaborations.

5.6.1 Instantiating well-formed groups

The parameters for well-formed groups are known and could likewise apply to B/CI cloudminds. Three important aspects are as follows: a central understanding of group dynamics is conveyed by the forming–storming–norming–performing heuristic (Tuckman 1965). This model of group development suggests that these phases are all necessary and inevitable for a team to grow, face challenges, find solutions to problems, plan work, and deliver results. The second aspect of well-formed groups is Simondon's group individuation principles. These principles stipulate that the group must reform with each new participant's entry or exit in order to vest all participants in the aims and responsibilities of the group (Swan 2015c). The third aspect of well-formed groups is an effective governance model.

One such governance model for organizing group tasks is Convergent Facilitation (Kashtan 2014). Transparency and voluntary participation in decision-making are the core principles, based on the idea that everyone wants to know how decisions are being made but may not necessarily be involved in all decisions. These three principles of forming–storming–norming–performing, group individuation, and transparent decision-making could provide an empowering beginning for the implementation of B/CI cloudminds.

The other implementation trajectory that is also necessary for peak cloudmind performance involves addressing potential barriers to large-scale group collaborations. These hindrances can be identified as the three C's of credit assignment, coordination (e.g. the sheer challenge of how to multithread human capabilities), and communication.

5.6.2 Overcoming barriers to large-scale group collaboration

5.6.2.1 Credit assignment

An initial barrier to effective team collaboration is credit assignment, which pertains to the issue of an individual securing credit for contributions to the group project, where the rewards (both extrinsic and intrinsic) match the contribution. For example, open-source software projects are considered situations of asymmetric efforts and rewards in that a substantial amount of value is created, which for the most part is unremunerated. The advent of blockchains assists with the resolution of the credit-assignment problem. All electronic activity is recorded, logged, and tracked, and an overlay of smart contracts (programmatic contracts) and machine learning can be applied to identify idea generation and reward value creation.

For instance, through online software versioning tools and databases such as GitHub, the capability now exists to track any line of open-source code used in any future project. The vast corporate codebases built on open-source software could be remunerated via an annuity-type mechanism with a royalty payment every time a line of software is called. A science fiction example of the idea of automated cloud-based tracking and remuneration of intellectual property creation as a standard feature of group collaboration contracts is outlined in *Rainbow's End* (Vinge 2007).

Through blockchain-based smart contracts and other digital tracking mechanisms, enough progress has been made toward solving the credit-assignment problem that individual minds (human and machine) might be comfortable participating in B/CI cloudminds. Blockchains provide a next-generation governance and social contracting tool for outlining the rights and responsibilities of B/CI cloudmind participation in granular detail (Swan 2019).

5.6.2.2 Coordination

Following credit assignment, the next potential barrier to cloudmind participation is the sheer practical challenge of multithreading participation into a coherent whole. As one task may have multiple threads to be knit together, so too do individual participants need to be brought together in a group endeavor. Assuming that the individuals agree to be coordinated in a collaborative effort, how does such a collaborative effort proceed? There are several existing examples of models for the effective multithreading of participation in group projects.

5.6.2.2.1 Agile programming

One premise is that small teams are most effective. At a minimum, the free-rider problem (uneven contribution as an unavoidable problem in team projects) is reduced, and accountability is high. An extremely pared-down model of small team formation is agile software programming. Two coders are paired, one who writes code and the other who reviews and debugs it, in tight interactive cycles at the smallest unit level of productivity. Agile pair partners for specific cloudmind tasks are a possibility, but this does not address the issue of how to combine hundreds and thousands of human and machine minds to address tasks of greater complexity.

5.6.2.2.2 Cathedral versus Bazaar

Further models of software development are the Cathedral and Bazaar models, realized as closed-source development and open-source development, respectively (for example, Microsoft versus Linux). The Cathedral relies on in-house developers who design, code, test, and publish proprietary software in significant releases. The Bazaar is a model where software is developed publicly on the internet in open-source projects with frequent releases of minor updates. Proponents of the open-source model claim that it is more efficient, in that bugs can be discovered more quickly with the source code being widely available for public scrutiny and testing, as opposed to the time and effort required to identify bugs in the closed model, where the code is available to only a few developers (Raymond 1999). There are inherent benefits for both the Cathedral and the Bazaar, and B/CI cloudminds could be modeled on either. Higher-risk endeavors often require a more hierarchical governance model (early warning signals of global financial contagion, for example), whereas creative endeavors tend to favor a flatter governance model (creative expression cloudminds).

5.6.2.2.3 Moon landing

A prime example of multithreading diverse human capabilities into a coordinated endeavor is the project management virtuosity of Mission Control, created at NASA by Christopher Kraft, Jr, in 1969. In the mission control concept, the key point is selecting a team in which each person is an expert. The core mission control team was about 45 people, with each desk in the iconic room having a specific focus (Kraft 2001). Each person was an expert and knew their particular systems more closely than anyone else. The greatest expert in each area was trusted, and responsibilities were clear.

The team was bound together in the overall mission, which was also clear and tangible, involving human life and the 'moonshot' vision. The five-control-floor Mission Control building at the Johnson Space Center in Houston, TX, continues to bear Kraft's name today. In terms of B/CI cloudmind design, the takeaway point is the division of labor. The same principle of enrolling only one expert on each specific topic could likewise be followed in cloudminds. The value of a galvanizing mission and charismatic leadership is also not to be underestimated.

5.6.2.2.4 Polymath project

Externalized (outside the body) cloudminds already exist in the form of collaborative online communities. An initial example is open-source software development. There are also wisdom-of-the-crowd group competition sites such as Kaggle and

R&D markets such as InnoCentive. One of the most interesting prototypes for B/CI cloudminds is the Polymath Project, which is an online collaboration to solve mathematical problems established in 2009. Several problems have indeed been solved, resulting in publications; for example, finding a combinatorial proof for the Hales–Jewett theorem (Polymath 2012), along with calculating bounds on the density of Hales–Jewett numbers (Polymath 2010).

Forty people contributed to solving the first problem, an effort that required seven weeks. A study of the project found that a small percentage of the users created most of the content; however, almost all participants contributed some content that was influential in resolving the problem (Cranshaw and Kittur 2011). The study also found that leadership played a key role and suggested design parameters for online collaboration communities regarding coordination, task identification, and the management of background materials.

5.6.2.3 Communication
Assuming that credit assignment and multithreaded coordination are resolved, the third barrier to effective group collaboration is communication. This work argues that misunderstanding may be seen as an interoperability issue between minds. Communication is a function of language and meaning, where language is the minimal barrier to entry for effective communication, as global collaborative teams already know. At a basic level, natural language translation is a solved problem due to the availability of Google Translate (which delivers a version of the Babel fish concept, an in-ear unit that performs simultaneous translation).The near-simultaneous translation of any electronically capturable medium is a real-life actuality and would likely be a standard feature in B/CIs. It is well known, however, that language only construes a small portion of meaning. Communication is purportedly 7% verbal and 93% nonverbal (55% body language and 38% voice tone) (Yaffe 2011). The problem is that language is a limited narrowband communication medium that leads to miscommunication and misunderstanding due to humanity's diverse value systems, cultural backgrounds, and experiences.With unprecedented secure access to the brain and biophysical responses (e.g. parasympathetic nervous system stimulation, neurotransmitter activation, etc.), there is an opportunity for B/CI cloudminds to innovate methods of beyond-language communication. There are many examples in science fiction; first and foremost is telepathy. Another possibility is direct neural transfer, contemplated in both science (Martins *et al* 2019) and science fiction (Chiang 2002). Nancy Kress elaborates on new concepts in group communication. One concept involves individuals experiencing 'head pain' as a signal of being out of alignment with the 'shared reality' of the collective unconscious (in the *Probability Moon* trilogy 2000–02). Another idea is the holostage described in the *Beggars in Spain* trilogy (1993–96). A holostage is a medium in which a thought string is projected in the form of a hologram with all of its related associations. The viewer can see the genesis and rationale for the thought (i.e. insights into the values of the thinker), thus reducing the possibility of miscommunication. Thought string projection is described as follows.

Speaking any single sentence to the computer, the holostage begins to form a three-dimensional shape of words, images, and symbols linked to each other. The holostage brings out the associations the mind makes, based on its store of past thought strings and algorithms for the way that mind thinks.

(Kress, 1994)

Instead of a single language-based sentence, a full thought is projected. B/CI cloudminds might likewise be based on a richer expression of participant thoughts, in projected holograms or other modes of wideband communication.

5.6.2.3.1 Interplanetary hash-linked data structure for the brain

B/CI cloudminds will exist in a digital environment. This means that not only can activity be seamlessly tracked and logged (for credit assignment and privacy protection), but also that other checks and balances can be applied, such as enforcing format compatibility. No transaction can enter the B/CI system without being in a compliant format. The benefit is that ideas might be brought into greater alignment from the beginning, based on the way they are presented.

The benefit of the digital format is that interoperability can be enforced. The Simple Mail Transfer Protocol (SMTP) forced interoperability between CompuServe and AOL, which were previously 'walled garden' properties, allowing only email communications between community members. Today, email can be sent to any recognizable address. Contemporary challenges such as 'big data is not smart data' are partially due to non-interoperable formats, for example, between different electronic medical record (EMR) systems. The B/CI cloudmind could force interoperability between thoughts, thereby partially reducing misunderstanding to a formatting problem.

InterPlanetary Linked Data (IPLD) is a general proposal for the interoperability of internet-based digital content (Benet 2017). In the IPLD system, each content item has an address, a URL-based location where it resides in the cloud, which can be called by other programs. The web is treated as a unified information space in which all content addressed in a compatible formatting system may be accessed. IPLD is an overlay standard for accessing such compatibly addressed data. A hash-linked data structure is employed to protect data security and privacy while providing the interoperable format. This means that the URL that contains the content is not given, but rather a hash of it. A hash is a fixed-length code that corresponds to the underlying content (the URL in this case), which can only be decoded by the hashing algorithm. Hash functionality is a known technology that is routinely used to securely send data such as passwords and credit card numbers across the internet. In the IPLD system, a hash of the URL location of addressable content is sent; the receiver then decodes it to access the content at that URL.

IPLD is essentially a file system for accessing the web's content, i.e. the vast corpora of all existing online content, for example, GitHub code and PubMed health publications. A similar content-addressing system could be used for the brain to implement both individual and group B/CI cloudminds. Just as IPLD is an overlay for the web, IPLD for the brain (IPLD/brain) could be an overlay for the B/CI.

If instantiated in a blockchain, B/CI entities and their transactions are already in IPLD-ready formats and can automatically participate in the IPLD content-addressing system. IPLD/brain is a specialized application of IPLD as a general data standard. For a B/CI cloudmind, any neuralnanorobot (axonal endoneurobot, synaptobot, or gliabot) and any of their transactional activities can be called in the IPLD/brain system. In the B/CI group cloudmind, one of the most relevant applications is interfacing thoughts to create new ideas and solve problems.

For the B/CI cloudmind, the IPLD/brain data structure first and foremost provides a secure content-addressing system in a cloud-based environment. The second benefit of the IPLD/brain data structure is enforced interoperability. Any digital content in the cloudmind brain system becomes interoperable. Just as an EMR blood pressure reading in the metric system is shareable due to the use of compatible formats, so too can ideas be better collaborated on with compatible formatting. Thoughts are digitally instantiated in the B/CI cloudmind system anyway and therefore can be easily interfaced with each other in team collaborations.

The overlay of the interoperable data structure protects the integrity of the under-lying content, in this case, the natural and original way that an individual mind thinks about and approaches a problem, y*et allows* the possibility of interoperating with the cloudmind so that ideas can be aligned at a formatting level for more effective collaboration. The aim of IPLD/brain is to reach peak-performance cloudminds more quickly by reducing basic understanding requests such as 'say more about that' or 'what do you mean?' Each idea in the B/CI cloudmind could have its own content-addressable location that is called in the hash-linked IPLD/brain data structure.

5.6.2.4 *The brain is a Merkle forest of ideas*

The brain is essentially a Merkle forest of ideas. A Merkle forest is a group of Merkle trees, each of which references an arbitrarily large thought trajectory and can be rendered digitally compatible through multihash protocols and Merkle roots. A Merkle root is a top-level hash that references an underlying data structure. For example, blockchains work in the structure of Merkle trees that roll up to a top-level Merkle root. One top-level Merkle root references the entire database of all Bitcoin blockchain transactions that have occurred since inception in January 2009. As of October 11, 2021, the Bitcoin data structure comprised over 704 000 transaction blocks (each with a few thousand transactions) (https://blockexplorer.com/). The entire database of transactions can be referenced by one hash code that is 64 characters in length (the Merkle root). Likewise, in a hash-linked data structure, there can be one top-level Merkle root that references the entire underlying data corpus, such as all GitHub code or all PubMed publications. A multihash protocol allows interoperability between different hashing algorithms.

In a data structure in which all content items are compatibly addressed, there is no limit to the total amount of data that can be rolled up and referenced with a short command. An entire brain can be addressed through a Merkle root, and likewise, a cloudmind. A cloudmind effectively becomes a Merkle forest of Merkle trees, all compatibly accessible through the hash-linked data structure of IPLD/brain. The Merkle root system of data organization allows an arbitrarily large data store to be referenced with a single hash code. Just as temperature is a short code that represents the thermodynamic situation of septillions of air particles circulating in a room, and the surface theory in the AdS/CFT correspondence encodes the messy interactions of the bulk volume in one fewer dimensions, the Merkle root likewise indexes an arbitrarily large data store.

Furthermore, the Merkle tree structure of hashes is not merely a content-addressing system that references an entire data structure with one short code; it is an active cryptographic security platform. Each access to the data structure performs a cryptographic check by rerunning the hash functions to confirm that the underlying data has not changed. The software automatically compares the top-level hashes of the hashed content in the data structure (transaction blocks in the case of Bitcoin) to confirm that there has been no change to the integrity of the underlying data, not even one bit at any location in the entire 627 000 block Bitcoin data structure, for example. The live cryptographic checking of the Merkle root functionality of data structures like IPLD/brain is crucial to ensuring the ongoing security of B/CI cloudminds and engendering participant trust.

IPLD allows the whole web (of compatibly addressed content), or in the case of IPLD/brain, a whole mind or a whole cloudmind, to be referenced with one Merkle root and checked for data integrity using real-time cryptographic security. The sweeping implication of IPLD is that there could be one top-level hash for the entire internet, which corresponds to all human knowledge, and likewise one top-level hash that corresponds to an entire brain.

Seeing a brain as a Merkle forest of ideas is a conceptualization paradigm for the implementation of individual and group B/CI cloudminds. In IPLD/brain, a brain becomes a Merkle forest of Merkle trees, namely a hash-linked data structure that can be connected through multihash protocols (to link different hashing algorithms and content trees in the Merkle forest of ideas). An entire brain area, brain, or group of brains in the B/CI cloudmind can be content-addressed in the IPLD/brain hash-linked data structure. IPLD/brain is precisely the kind of secure, interoperable, scalable infrastructure needed to enact the B/CI cloudmind.

It has been proposed that thinking could be instantiated in a blockchain, in the brain-as-a-DAC concept (DAC: decentralized autonomous corporation; a collection of automated smart contract programs) (Swan 2015b). IPLD/brain is essentially the Brain DAC II, in the sense of being able to realize B/CI group cloudminds. The advance provided by IPLD/brain is that numerous brains can be multithreaded for collaborative endeavors. One initial task of IPLD/brain might be attempting to log the estimated 60 000 thoughts per day that each mind has (Al-Ghaili 2017) and identifying which might be most useful for novel ideation.

5.6.3 Cloudmind activities: what does the brain/cloud interface cloudmind do?

Assuming that B/CI cloudminds can be made possible, the question arises as to what they will do. The immediate objective of the B/CI is to map, monitor, cure, and enhance neural activity. At the more abstract level of everyday activities, the purpose of the B/CI is to facilitate human productivity, well-being, and enjoyment,

Table 5.19. Maslow's hierarchy of needs and beyond.

Maslow's tiers	Objective	B/CI measure
Maslow 1	Physiological survival	Energy, glucose, oxygen, ATP
Maslow 2	Psychological well-being	Neurotransmitter balances
Maslow 3	Self-actualization	Ideas, neurotransmitters, energy
Beyond Maslow	New levels of achievement	Ideas

which could be measured by the extent to which Maslow's hierarchy of needs is fulfilled (table 5.19).

Maslow's hierarchy of needs is a tiered structure that comprehensively articulates the different levels of human needs (McLeod 2007). Maslow identifies multiple levels that are consolidated here into three tiers as pictured. Maslow 1 includes the physiological survival needs for food, water, warmth, sleep, sex, and security. Maslow 2 contains the psychological needs for belonging, acknowledgment, and love. Maslow 3 denotes the self-actualization needs for achievement, creativity, and the realization of one's potential. A hierarchy is implied, in that lower-level needs must be met before progressing to the next tier. Pursuing interesting projects (Maslow 3: self-actualization) makes little sense if one is worried about food and housing for the night (Maslow 1: physiological survival). B/CIs could track the fulfillment of Maslow's hierarchy of needs (physiological survival, psychological well-being, and self-actualization) via biological cues logged by the B/CI and formalized in the notion of Maslow smart contracts (Swan 2019).

Since the cloudmind may be a collaboration of human and machine minds, its capacities could progress beyond those articulated in Maslow's hierarchy of needs, as contemplated in the beyond-Maslow category. Some of the potential areas of beyond-Maslow development could include scientific advances, cloudmind learning, and the cloudmind itself as a platform. In the area of scientific advances, cloudminds might pursue different strategies for integrating research findings, making them more broadly accessible to address the challenge of reductionist findings and the practical impossibility of keeping up with scientific publications. There could be journal club cloudminds that perform literature summaries, write review papers, and generally engage in knowledge-stewarding activities as next-generation librarians or information scientists. These kinds of integrative activities could themselves result in new findings (Chiang 2002).

Regarding learning, the B/CI is envisioned as a heightened learning tool that might support beyond-Maslow objectives, including through the eventual direct neural transfer of information. All the world's knowledge might be modularized for uptake and propagation. *Knowledge modules* may emerge as standardized units of information required for the understanding of a certain topic, packaged into consumable units, and distributable globally. Coursera (and other massive open online courses (MOOCs)) are platforms that deliver knowledge modules. The most successful (widely propagated) knowledge module (and cloudmind community prototype) is Andrew Ng's machine learning course, with over 3.5 million enrolled students as of April 2020 (Ng 2025).

Just as there are fungible global equivalents for healthcare (Swan, forthcoming), there could be fungible equivalents for learning, implemented via B/CI. Knowledge modules could be organized into structured paths for mastery (accredited degrees and certifications), deploying the same kind of leveling-up gamification techniques that have proven successful in learning communities as diverse as Toastmasters and World of Warcraft.

Considering the cloudmind itself as a platform, the beyond-Maslow objective would be to design and test new versions of intelligence with cloudminds. B/CI cloudminds could be used to instantiate different proposed models of artificial intelligence (AI). For example, one idea is to boot up Kant and Hegel's models of consciousness and various AIs envisioned in science fiction (for example, the multiparty society articulated by Max Harms 2016). With this recursive objective, the cloudmind could help produce new and better cloudminds.

5.6.4 Classes of cloudminds

A variety of classes of cloudminds may evolve, with different purposes (table 5.20).

Table 5.20. Different possible classes of cloudminds.

Classes of cloudminds
1 **Question-asking cloudminds:** question-asking cloudminds focus on generating good questions. Contemplating the human–machine partnership, futurist Kevin Kelly notes that humans are good at asking questions, and machines are good at answering them. (Neurocurrency: questions.)
2 **Problem-solving cloudminds:** working on a list of unsolved problems in various fields, problem-solving cloudminds might be a workhorse class of cloudminds, consolidating wisdom-of-the-crowd talent to solve the thorniest problems available. This could lead to the polymath Fields Medal mathnet, quantum gravity, and anti-aging cellular therapies cloudminds. Various methods might be applied (deduction, induction, abduction). (Neurocurrency: ideas, solutions, publications.)
3 **Journal club/affinity (special interest) cloudminds:** any variety of special interest cloudminds might focus on topics such as space exploration, transnational political systems, environmental sustainability, longevity, economics, painting, and music theory. (Neurocurrency: ideas, publications.)
4 **Transparent shadowing cloudminds:** transparent shadowing cloudminds could be a class of cloudminds devoted to the real-time immersive experience of other humans (on a permissioned basis), experiencing another person's life through their own eyes for learning, collaborating, apprenticing, and sharing experiences. (Neurocurrency: learning, experience, empathy units.)
5 **Random cloudminds:** variety and surprise are crucial, and the random cloudminds could be a serendipity-class cloudmind. This is reminiscent of the virtual entertainment device (stimsim) in *Neuromancer* (Gibson 1984), a ~250 channel device (i.e. a lot at the time), including a channel for those who do not know what to select. (Neurocurrency: enjoyment units.)

5.7 Risks and limitations

The biggest potential risk and limitation of the B/CI is security. There is 'no neural dust without neural trust.' Security plays an outsized role in the implementation of the B/CI because the brain is an extremely sensitive area for intervention. Given this sensitivity, the knee-jerk reaction is to reject any type of brain-connected internet device, much less onboard nanorobots, and the monitoring, sharing, and manipulation of neural information. Particularly in an era of heightened internet privacy and cybersecurity threats, the last thing that seems to make sense is interfacing the most sensitive asset humans possess, their brain, to the internet. Therefore, cybersecurity concerns must be resolved in a robust manner, or the B/CI will be a nonstarter.

One indication of the sensitivity of the brain is how individuals respond to information about their genetic profile for Alzheimer's disease. A landmark study, the REVEAL study, shared genomic information regarding Alzheimer's disease with individuals (Green *et al* 2009). Although notoriously intransigent to health-related behavioral modification, over half of the individuals finding out that they were at high risk for Alzheimer's disease changed their behavior as a result, using medication, vitamins, diet, and exercise (Chao *et al* 2008). Likewise, genomic sequencing pioneer Craig Venter engaged in interventional activities upon learning that his own genome (which he had sequenced) was positive for the Alzheimer's disease-related APOE 4 mutation (Nave 2016).

This work therefore proposes B/CI neural trust as a new concept in internet security, which would be adequate for the realization of the B/CI. B/CI neural trust provides a specific solution to deliver neural trust in the B/CI that is both hardware and software-based. Neural trust involves hardware trust conferred by nature's quantum information features and software trust provided by the blockchain operating software. Trust is provided in the quantum computing hardware solution for the B/CI through nature's built-in security features that apply to any quantum information domain, such as the no-copying and no-measurement principles. Trust is provided by the operating software proposed for the B/CI through the security and privacy features built into blockchains.

Beyond security concerns, the most substantial risk attached to this work is that the B/CI currently has the status of being a speculative proposal without immediate practical development possibilities. Only the core BCI exists for basic computerized cursor control and neuroprosthetic administration. However, the point is not to dismiss this work as science fiction or navel-gazing, since it is important to think through potential future technologies involving the brain in great detail. The B/CI and B/CI cloudmind are technologically complex concepts, and many aspects of how they are practically interfaced with human brains will need to be considered. It is crucial to consider these kinds of future technologies in detail, including various scenarios, design specifications, and implementation plans, before their actual advent.

Considering potential technologies ahead of their implementation could facilitate the development of social maturity and responsible technology use, which often lags behind the invention of a technology. Furthermore, considering sophisticated projects such as the B/CI and the B/CI cloudmind highlights what may be possible in the near term; for example, predictive data analysis of existing BCIs that relates to power consumption and component failure rates.

A second class of risks pertains specifically to the proposals in this work, namely that they may be ill-founded, infeasible, or inaccurate. Admittedly, this is a possible risk; however, the overall value of this work may lie in the gist of its theoretical ideas. It is expected that technical details may shift in the potential realization of B/CIs. Quantitative estimates of brain activity are an early-stage analysis and may vary widely in actuality. Also, given the sensitivity of the human brain, it may never be deemed appropriate to intervene, or at least not for a very long time (direct genomic intervention, for example, has been curtailed due to early missteps involving the death of a clinical trial participant in 1999 (Branca 2005)). The theoretical analysis proposed here may prove faulty in different ways.

For example, a more straightforward control theory not based on the AdS/CFT correspondence may may emerge for managing the interface between macroscale reality and quantum mechanical domains if quantum computing were to proceed more substantially. Any number of software-based control theories might be indicated. Blockchains are complicated to understand and might not survive, although large-scale autonomously operating global smart network computing software might continue to use some of the same core features.

A third class of risks pertains to the asymmetric treatment of opportunity and threat in this analysis. Many positive possibilities for the B/CI and the B/CI cloudmind are presented, but a corresponding and equally weighted consideration of the potential risks is not elaborated or resolved. Any technology is dangerous to the extent that the brain is involved. In an era of routine cybersecurity attacks and the inability of institutions and individuals to protect electronic data on personal computers and smartphones, it is naïve to think that anyone would provide internet access to their mind. In response, the B/CI and B/CI cloudmind are future-class technologies, and it is too early in the development cycle for a comprehensive risk assessment. Further, it should be noted that the B/CI will be a nonstarter without an adequate security solution. Future work necessary for B/CI implementation would certainly involve a more comprehensive risk analysis. Meanwhile, there are some science fiction examples of scenarios to avoid in B/CI design regarding the disruptive effects of neural hacking (for example, Bear 1998).

A final class of risks relates to the sequential arrival of new technologies. Predictions involving future technologies are notoriously difficult. Further compounding predictions related to any one technology is the issue of technology sequencing, meaning the order of arrival of different potential future technologies. The B/CI might be obviated by the advent of unforeseen technologies. For example, if noninvasive personal connectome mapping were to become possible by a non-B/CI means, there could be digital copies of each person's brain. This could make biological intervention much less desirable and supersede the biological brain with the digital version as the desired platform for all manner of enhancement and collaboration. In this case, the human brain might be replatformed to the digital environment without the need for B/CIs.

5.8 Conclusions

The full suite of BCI technologies includes the core, existing BCI (brain–computer interface) for directing computer cursors and neuroprosthetics, and the envisioned cloudmind B/CI for safely connecting one or more minds to the internet cloud. The B/CI would be implemented through an ecosystem of micron-scale medical nano-robots seamlessly embedded within the brain. The neural signaling of the brain would be platformed in an adjacent computational environment to simulate and enhance the brain's native activity. At least three neuralnanorobot species have been outlined to comprise the B/CI. These are axonal endoneurorobots corresponding to the axons that generate electrical pulses (action potentials) in neural signaling, synapto-bots that reside within and around the synaptic terminals that transmit and receive signals, and gliabots linked to the glial cells that facilitate signal transmission and cleanup. There would be a one-to-one correspondence between neural cells and neuralnanorobots, translating to a neuralnanorobot complement for each of the brain's 86 billion neurons, 85 billion glial cells, and 242 trillion synapses.

The technical requirements for the realization of the B/CI are substantial. This work proposes quantum computing as the hardware platform for the B/CI (including via the possible advent of global quantum photonic networks), together with a holographic control theory (based on the AdS/CFT correspondence) as the lever for the macroscale control of the quantum computing cloud environment (the AdS/CFT correspondence is a universal control theory that could be used to orchestrate macroscale-quantum domains), and a bioblockchain neuroeconomy as the operating software of the in-brain B/CI neuralnanorobot network. Security is a paramount concern for B/CI implementation, and neural trust is provided first as hardware trust from nature's built-in quantum security features (such as the no-cloning and no-measurement principles) and second as software trust from the cryptographic properties of blockchains.

Some of the additional benefits of instantiating the neural signaling platform in the digital environment can be seen in the greater implementation of group cloudmind B/CIs for interactive collaborations to enhance productivity, well-being, and enjoyment. Lower-level barriers to group collaboration can be overcome, such as credit assignment and the sheer feasibility of multithreading thousands of minds into a coherent whole. A richer level of communication may be made possible by B/CIs, such as direct neural transfer. In the digital environment of coded thoughts, formatting standards can enforce a basic level of interoperability that might reduce the possibility of misunderstanding and improve the ability to collaborate on ideas.

On one hand, the B/CI is required merely to cope with modern reality, in which the development of science and technology clearly outpaces that of biology. Beyond that, the greater potential of the cloudmind B/CI for group collaboration (the 'ten-billion-synapse world mind') is the possibility of making progress toward the achievement of a Kardashev-plus society that is able to marshal all available tangible and intangible resources by mental and physical means to enable a flourishing society.

Glossary

AdS/B/CI (quantum B/CI): AdS/CFT correspondence model of B/CI technologies (BciTech)

AdS/brain: multitier AdS/CFT correspondence neural signaling model

AdS/CFT correspondence: the AdS/CFT (Anti-de Sitter space/conformal (basic) field theory) correspondence (also called the holographic correspondence, gauge-gravity duality, and the bulk–boundary relation) is the claim that in any physical system, there is a correspondence between a volume of space and its boundary region, such that the interior bulk can be described by a boundary theory in one fewer dimensions.

BCI (brain–computer interface): A BCI (brain–computer interface) is a direct communication pathway between a wired brain and an external device achieved via EEG (electroencephalography).

B/CI (human brain/cloud interface): a B/CI is a proposed technology that would interface the human brain with the internet cloud using neuralnanorobots.

B/CI cloudmind: A B/CI cloudmind is one or more minds connected to the internet cloud on an individual or group basis for activities related to productivity, well-being, and enjoyment.

B/CI technologies: B/CI technologies are the suite of technologies, including the core B/CI and the individual and group cloudmind B/CI, that safely connect the mind to the internet.

Bioblockchain: a bioblockchain is a blockchain implemented in the biological context.

Black hole information paradox: the black hole information paradox asks how it is possible for information to exit from a black hole, since information cannot escape from black holes (per general relativity), but Hawking radiation does emanate from black holes, which conflicts with the no-cloning theorem of quantum information (per quantum mechanics).

Blockchain (distributed ledger technology): a blockchain (distributed ledger technology) is a secure automated cryptographic system for internet-based value transfer.

Cloudmind: A cloudmind is one or more minds connected to the internet cloud. A cloudmind might be comprised of an individual mind operating on the internet or multiple human and machine minds participating in a collaborative activity. 'Mind' generally denotes an entity with processing capacity (not necessarily a biological mind that is conscious).

Complex adaptive systems: complex adaptive systems are those that are nonlinear, emergent, open, unknowable at the outset, interdependent, and self-organizing.

ComplexityTech: A ComplexityTech (complexity technology) is a technology for managing complexity and complex adaptive systems whose behaviors are nonlinear and unpredictable.

Computational complexity: computational complexity is a schema of the tiers of the necessary computational resources (in time and space) needed to calculate a given problem.

Connectome: the connectome is a wiring diagram of the brain showing the functional processing framework and structure of the brain in full spatial and temporal resolution.

Gauge-gravity duality: gauge-gravity duality is an interpretation of the AdS/CFT correspondence based on the intuition that both gravity (general relativity) and gauge theory (quantum mechanics) are field-based phenomena at the scale of gauge theory (1×10^{-15} m) and therefore might be joined through the correspondence.

Hash code: a hash code is a function used to map data of arbitrary size onto data of fixed size.

Holographic principle: the holographic principle is the notion of reconstructing a 3D volume on a 2D surface (like a hologram) and indicates two valid views of the same phenomenon.

IPLD (InterPlanetary Linked Data): IPLD is an internet-wide file system overlay for access to compatibly addressed content; a data standard for referencing any variety of the vast corpora of internet-based data structures such as GitHub code and PubMed health publications. URL links are hashed for data security and to provide the interoperable format.

IPLD for the brain: IPLD/brain is a specialized application of IPLD as a B/CI data standard.

Kardashev-plus society: a Kardashev-plus society is one that is able to marshal all tangible and intangible resources, including energy, for the survival and success of populations, as measured by productivity, well-being, and enjoyment.

Kardashev civilizations: the Kardashev civilization scale is a scale that measures societal advancement using the degree of energy resources controlled. A Type I civilization is able to use all available sunlight on a planet, Type II all the energy the sun produces, and Type III the energy of the entire galaxy. Humanity is currently estimated to be at 0.7 on this scale.

Knowledge module: a knowledge module is a standardized unit of information required for the understanding of a certain topic, packaged into a consumable unit and globally distributable.

Medical nanorobots: medical nanorobots are nanorobots (molecular machines at the nanoscale (1×10^{-9} m)) designed to complement native cells and perform medically related tasks in the body. Proposed species include respirocytes, clotto-cytes, vasculoids, and microbivores.

Merkle forest: a Merkle forest is a group of Merkle trees referencing multiple content trajectories.

Merkle root: a Merkle root is a top-level hash (e.g. 64-character short code) that references an entire underlying data structure (Merkle tree), for example, an entire database of transactions.

Neurocurrency: a neurocurrency is a resource (e.g. ions) used to execute a neural function.

Neuroeconomy: the neuroeconomy is a multiagent economic system used to operate neural activities based on goal-directed behavior in transactions denominated in neurocurrencies.

Neuron: a neuron is an electrically excitable cell that communicates with other cells by sending a signal called an action potential to other neurons across specialized connections called synapses. A neuron is comprised of a cell body (soma), a long thin axon insulated by a myelin sheath for outbound signaling, and multiple dendrites for receiving inbound signals.

Neuralnanorobots: neuralnanorobots are medical nanorobots designed specifically for operating in the brain. Three species of neuralnanorobots have been proposed to correspond to the phases of the neural signaling process: axonal endoneurobots, synaptobots, and gliabots.

Payment channel: a payment channel is a network overlay that allows parties to precontract with one another to automatically replenish resources or rebalance accounts.

Quantum B/CI: B/CI realized via quantum computing (superposition, entanglement properties).

Quantum computing: quantum computing is information processing at the quantum scale of atoms (1×10^{-9} m), particularly computing based on SEI properties (superposition, entanglement, interference).

Quantum internet: the quantum internet is a next-generation internet proposal that would feature secure end-to-end communication based on quantum switches and routers, quantum key distribution, quantum processors, and quantum memory.

SEI properties: the SEI properties (superposition, entanglement, and interference) are the properties of quantum objects (atoms, ions, photons) that facilitate quantum computing.

Smart contract: a smart contract is blockchain-based pre-specified programmatic instructions.

Smart network: smart networks are automated global network-based computational technologies, such as blockchains, deep learning networks, UAVs, and automated trading networks.

Synapse: a synapse is a structure that permits a neuron to pass an electrical or chemical signal, mainly to another neuron, or possibly to another cell or the intercellular environment.

Temperature term: a temperature term is an aggregate informational state of a system that might be employed as a control lever. Examples: Merkle root, AdS/CFT surface theory.

Zero-knowledge proof: A zero-knowledge proof is a proof that reveals no information except the correctness of the statement. Data verification is separated from the data itself, conveying zero knowledge about the underlying data, thereby keeping it private.

Appendix A Relative sizes of neural entities and neuralnanorobots

This analysis suggests that the axonal endoneurobot and the gliabot could be similar in size to other medical nanorobots, about 1000 nm or larger; however, the synaptobot would need to be much smaller, perhaps on the order of 30–300 nm, given the small size of the synaptic area where it is to be housed. Several species of medical nanorobots have been proposed for health-related activities in the body (generally 1000–3000 nm in size), as listed in table 5.A1-1.

Table 5.A1-1. Relative sizes of circulatory system entities and medical nanorobots.

Entity	Size (microns)[5]	Size (nm)	References
Human hair and circulatory system			
Human hair	100[6]	100 000	Ley (1999)
Red blood cell	7	7000	Freitas (2012, p 69)
Smallest capillaries	3	3000	Freitas (2012, p 69)
Medical nanorobots			
Clottocytes (artificial platelets)	2	2000	Freitas (2000)
Microbivores (artificial phagocytes)	3.4	3400	Freitas (2005)
Respirocytes (artificial red blood cells)	2–3	2000–3000	Freitas (2012, p 69)
Vasculoids (cell transporter boxcar)	100 × 6	100 000 × 6000	Freitas and Phoenix (2002)
Nanorobot components		1–10	Freitas (2012, p 69)
Vascular cartographic scanning nanodevice (for connectome mapping)	1	1000	Domschke and Boehm (2017)

A similar analysis can be performed for neuralnanorobots. The objective is to clarify the implied size of neuralnanorobots, given the size of the neural areas in which they are to be housed. The relevant neural cells are listed in table 5.A1-2. Whereas there is a lot of room in the neuronal cell body at the start of the axon, where the axonal endoneurobot is to be located, and in the glial cells for the gliabot, there is very little space available in the synaptic terminal areas for the synaptobot. There is room for a 1000 nm nanorobot in the 10 000–25 000 nm neural cell body and the 15 000–30 000 nm glial cell. However, the synaptic terminal areas are only 100–1000 nm^3 in size, so an embedded synaptobot would need to be perhaps 30–300 nm in size.

The precise sizes and functions of various areas involved in presynaptic and postsynaptic signaling terminals constitute an active area of research. The overall synapse is comprised of the presynaptic terminal, the synaptic cleft, and the postsynaptic terminal. Although some presynaptic terminals are micron-sized (1000 nm^3) (table 5.A1.2(A), 5.A1.2(B)), the vast majority are less than 0.1 μm (less than 100 nm^3) (table 5.A1.2(C)) (Kleinfeld *et al* 2011). The synaptic cleft is 20–50 nm (Scimemi and Beato, 2009). Synaptic vesicles in the terminal area occur in two varieties, either large (200 nm^3) or small (50 nm^3) (https://web.williams.edu/imput/IIA1_right.html); they are 40 nm^3 in the visual cortex, for example. A particular apparatus called the postsynaptic density is a protein-dense area attached to the

---

[5] Regarding units: 1 micron = 1 micrometer = 1 μm. A micron is a standard unit of length equaling 1 x 10^{-6} meter (SI standard prefix 'micro-' = 10^{-6}) for one millionth of a meter (or one thousandth of a millimeter, 0.001 mm) which is equal to 1,000 nanometers. A nanometer is one billionth of a meter (1 x 10^{-9} meter).
[6] Range: 17–181 μm.

Table 5.A1-2. Relative sizes of neural cells. (Table created by the author.)

Neural cells	Size (microns)	Size (nm)	References
Neuron cell body (soma)	10–25	10 000–25 000	Chudler (2009)
Neuron cell body nucleus	3–18	3000–18 000	Chudler (2009)
Presynaptic and postsynaptic area	0.1–1 μm^3	100–1000^3	Kleinfeld *et al* (2011)
Synaptic cleft		20–50	Scimemi and Beato (2009)
Glial cells: astrocytes	40–50	40 000–50 000	Parent (1996)
Glial cells: microglial cells	15–30	15 000–30 000	Kettenmann and Verkhratsky (2011)

Table 5.A1-3. Implied sizes of neuralnanorobots dictated by the sizes of the locations housing them. (Table created by the author.)

Neuralnanorobot	Location	Size of location (nm)	Size of neuralnanorobot (nm)
Axonal endoneurobot	Neuron cell body (soma)	10 000–25 000	1000
Synaptobot: presynaptic terminal	Presynaptic terminal	100–1000	30–300
Synaptobot: synaptic cleft	Synaptic cleft	20–50	5–10
Synaptobot: postsynaptic terminal	Postsynaptic terminal	100–1000	30–300
Gliabot	Glial cells	15 000–30 000	1000

postsynaptic membrane, which is from 250–500 nm in diameter (Meyer *et al* 2014). The postsynaptic density is located across from the active zone of the presynaptic terminal and orchestrates signal reception, becoming enlarged during synaptic plasticity (the long-term potentiation or depression of synapses). In general, the sizes of presynaptic and postsynaptic terminals are dynamic, as they expand and shrink during the signaling process. The signaling of many different proteins is implicated in the process. For example, Sonic hedgehog (Shh) signaling has an expansionary effect on the presynaptic terminals of both glutamatergic and GABAergic synapses in adult hippocampal neurons (Mitchell *et al* 2012). Glutamate (excitatory action) and GABA (inhibitory action) are two of the most common neurotransmitters in the brain, comprising 90% of neurotransmitter activity.

Finalizing the neuralnanorobot size analysis, the axonal endoneurobot and the gliabot could be analogous in size to other nanorobots, at ~1000 nm in diameter (table 5.A1-3).

Table 5.A1-4. Neurotransmitter concentrations (millimoles (mM)) in synaptic vesicles. (Table created by the author).

Neurotransmitter	Concentration (mM)	References
Acetylcholine	0.3–260 mM	Scimemi Beato (2009)
Glutamate	60–150 mM	Scimemi Beato (2009)

Table 5.A1-5. Neurons by type and function. (Table created by the author.)

Neurons	Function
Motor neurons	Control the activity of muscles and higher-level cognitive functions such as reasoning and language (most numerous by far).
Sensory neurons	Detect and respond to internal and external stimuli.
Interneurons	Moderate reflexes and coordinate activity between neurons.

The implied size for the synaptobot, though, is much smaller: 30–300 nm in diameter if housed within the synaptic terminals and smaller still if located at the synaptic cleft, perhaps 5–10 nm in diameter (the size of a nanorobot part).

Determining neurotransmitter concentrations in synaptic vesicles has been a helpful first step in quantifying the size and scope of the activities involved in synaptic signaling. The estimated concentrations of the primary neurotransmitters acetylcholine and glutamate appear in table 5.A1-4.

As the background for neuralnanorobot size analysis, neurons are the main component of nervous tissue; they are electrically excitable cells that signal each other through synapses. Based on function, there are three primary types of neurons (table 5.A1-5). Motor neurons are responsible for motor control, which is the planning and coordination of movement, as well as reasoning, language, and higher-level cognitive functioning. Sensory neurons receive and process sensory information, including visual stimuli. Interneurons connect neurons to other neurons within the same brain region to coordinate activity between groups of neurons.

Glial cell processing takes place on the very small scale of subcellular processing. However, glial cells may be ideally poised for intervention, given their extensively catalogued operations and locational adjacency to neurons. It may be easier and safer to intervene with glial cells as the support environment for neurons rather than directly with the neurons themselves. The different types of glial cells in the central and peripheral nervous system are listed in table 5.A1-6.

The suggested uses of glial cells in the B/CI design are as follows. Oligodendrocytes generate the myelin sheath to protect axons and are in close proximity to axons. Astrocytes are active around the synaptic cleft in facilitating cell signaling. Microglia serve as a localized immune system to identify and destroy pathogens and could be used in B/CI projects to target neurodegenerative diseases. Microglial interventions

Table 5.A1-6. Glial cells by type, frequency, and function. (Table created by the author.)

Glial cells	Percentage	Function
Central nervous system		
Oligodendrocytes	45%–75%	Provide myelination to coat and insulate axons.
Astrocytes	19%–40%	Regulate chemical environment by removing excess potassium ions and recycling neurotransmitters: ATP, IP3 (messenger enzymes), calcium.
Microglia	<10%	Destroy pathogens and phagocytose debris; microglia deficiency is associated with Alzheimer's disease, Parkinson's disease, and amyotrophic lateral sclerosis (ALS).
Ependymal cells		Secrete cerebrospinal fluid and produce the blood–brain barrier.
Radial glia		Neuroepithelial development and neurogenesis.
Peripheral nervous system		
Schwann cells		Similarly to oligodendrocytes, they provide myelination to axons in the peripheral nervous system, phagocytose to remove debris.
Satellite cells		Surround neurons in sensory, sympathetic, and parasympathetic ganglia; regulate external chemical environment; respond to ATP and calcium ions.
Enteric cells		Provide homeostasis and muscular digestive processes; located in the intrinsic ganglia of the digestive system.

Table 5.A1-7. Neuron-to-glia ratios by brain area. (Table created by the author.)

Brain area	Neuron-to-glia ratio	Glia (bn)	Neurons (bn)
Cerebral cortex			
Gray matter and white matter	3.72	60.84	16.34
Gray matter	1.48		
Cerebellum	0.23	16.06	69.03
Basal ganglia, diencephalon, brain stem	11.35		

might be employed to address Parkinson's disease, Alzheimer's disease, and amyotrophic lateral sclerosis (ALS). The overall ratio of neurons to glial cells is estimated to be about one to one (86 billion neurons and 85 billion glia). However, there is a great deal of specificity, depending on the brain region (Azevedo *et al* 2009). In the cerebral cortex, the gray matter neuron-to-glial-cell ratio is thought to be 1.48 (while the ratio in the overall cerebral cortex is 3.72). The same ratio in the cerebellum is only 0.23 (relatively few glial cells). The basal ganglia, diencephalon, and brain stem region, on the other hand, have a ratio of 11.35 to one (table 5.A1-7).

Appendix B B/CI technical requirements and implementation phases

This appendix discusses some of the prominent technical requirements for B/CI implementation, such as bandwidth and information transfer, transaction processing, and power, and outlines a phased implementation plan. A framing specification of the technical requirements for a long-term, nondestructive, real-time human brain interface with the cloud appears in table 5.A2-1 (Martins *et al* 2019).

Bandwidth and information transfer

The B/CI neuralnanorobots would need an extremely fast wireless transmission capacity, on the order of 6×1016 bits s^{-1} (Martins *et al* 2019). The idea is to transmit synaptically processed and encoded human brain electrical information via auxiliary nanorobotic fiber optics (30 cm^3) with the capacity to handle up to 10^{18} bits s^{-1} and provide rapid data transfer to a cloud-based computing environment for real-time brain-state monitoring and data extraction (Martins *et al* 2019).

To some degree, existing BCIs for controlling neuroprosthetics through electrical pulse brain signals are offering internet connectivity. For example, Hanger, a Texas-based provider of orthotic aids and prosthetic limbs, collects near-real-time data regarding usage and mobility, connecting directly to AT&T's 4G Long-Term Evolution for Machines (LTE-M) network (Scroxton 2018). Shih *et al* (2012) called for personal data protection standards to be incorporated into the design of BCI medical applications, particularly in signal-acquisition hardware and software, which should be convenient, portable, safe, and able to function in all environments.

It is estimated that full-fledged B/CIs will require broadband access with extremely high upload and download speeds compared to today's rates. Internet networks are starting to accommodate two-way transfer. Although initially designed for asymmetric information downloads from servers to clients, communications networks now support large volumes of data uploads from IoT sensors and consumer devices. The current data rate for a B/CI-type upload is 24 Mb s^{-1}, which can be supported by both Bluetooth 4.0 (for IoT) and IEEE 802.11n low-power Wi-Fi technology (for body area networks (BANs)) (Zao *et al* 2014). Next-generation communications networks, such as 5G (100–200 MB download speeds) and farther-future terahertz networks (100 GB data links), may play a role in the ultrahigh-speed wireless data networks of the future that could enable the required upload and download speeds for B/CIs.

Table 5.A2-1. B/CI technical requirements. (Table created by the author.)

	Neuralnanorobot requirements
1	Ultrahigh-resolution mobility.
2	Autonomous or semiautonomous activities.
3	Nonintrusive ingress/egress into/from the human body.
4	Robust information transfer bandwidth for interfacing with external computing systems.

Intrinsic neural firing rates

There is sparse information available regarding the intrinsic neural firing rates in humans that could be utilized as an input in calculating the transaction-processing capacity required by the B/CI. A widely employed technique is to estimate the rate of firing in the human neocortex based on the brain's energy budget (in other words, back-calculating the implied firing capacity). Results vary, as one analysis estimates four spikes per second for human neocortical firing (Attwell and Laughlin 2001), while another suggests an average firing rate of $\sim$0.16 times per second with an overall range of 0.3–1.8 times per second (AI Impacts 2020). Regarding other species, one study of the visual cortex found rates of neural firing averaging three to four spikes per second in cats and 14–18 spikes per second in macaques, while another found nine spikes per second in cats (Baddeley *et al* 1997). Other work in the neocortical simulation efforts of the whole-brain emulation project has noted the difficulty of estimating an average firing rate due to the nonlinear behavior of one neuron triggering another (Gerstner *et al* 1997). Another variable to consider in neural signaling is the necessary refractory period, which is a few milliseconds in humans (Nicholls *et al* 2012).

The world's largest state-of-the-art transaction-processing systems accommodate 175 000 transactions per second (table 5.A2-2). This implies that if neurons were firing once per second (although the real figure might be higher), then the firing of $\sim$500 000 (0.001%) of the brain's 86 billion neurons could be handled by contemporary transaction systems.

In-situ power

A key design requirement for neuralnanorobots is the ability to sustain themselves by foraging for fuel in the local biological environment. Estimates suggest that nanorobots of the standard estimated size (1000 nm in diameter) could produce several tens of picowatts of power from oxygen reaching their surface naturally in the blood plasma (Hogg Freitas 2009). This would provide enough power for the steady-state activities of the nanorobot. If further equipped with pumps and tanks for onboard oxygen storage, nanorobots could possibly collect enough oxygen to support burst power demands two to three orders of magnitude larger. Software

Table 5.A2-2. Selected contemporary transaction systems and their transactions-per-second (TPS) rates. (Table created by the author.)

	Transaction system	Average TPS	Peak TPS	Year	Reference
1	Visa	2000	24 000	2011	Visa (2011)
2	Alipay (China)	120 000	175 000	2017	Skinner (2017)
3	Facebook	175 000	N/A	2017	Ehrsam (2017)
4	World's largest banks	100 000	N/A	2020	Blaschka (2020)

Table 5.A2-3. B/CI implementation phases by functionality. (Table created by the author.)

B/CI function	Phase I	Phase II
Scope of signaling	Electrical	Electrical and chemical
	Action potentials: capture electrical signal processing	Neurotransmitters: capture electrical and chemical signal processing
Temporal processing	Batch upload	Real-time processing
Information transfer	One way (outbound)	Two-way information retrieval
Scale	Cellular processing: neurons; atomic scale (1×10^{-9} m)	Subcellular processing: synapses; gauge theory scale (1×10^{-15} m)
Focus	Targeted brain regions	Whole brain

updates and lifecycle management are likewise design considerations that could hopefully be instituted for autonomous operation.

B/CI implementation

The B/CI project would be implemented in phases (table 5.A2-3). A general list of the basic functionality in the minimal implementation is highlighted in the Phase I column, and the extended, full-fledged version of the functionality is highlighted in the Phase II column. Although neural signaling is both electrical and chemical, the initial B/CI implementation may be possible on the basis of electrical signaling alone (Martins *et al* 2019).

References and further reading

AI Impacts 2020 Neuron firing rates in humans https://aiimpacts.org/rate-of-neuron-firing/ (accessed 6 May 2020)

Al-Ghaili H 2017 The human body in number *YouTube* https://youtube.com/watch?v=Lcu8fz2jXr4 (accessed 6 May 2020)

Attwell D and Laughlin S B 2001 An energy budget for signaling in the grey matter of the brain *J. Cereb. Blood Flow Metab.* **21** 1133–45

Azevedo F A, Carvalho L R, Grinberg L T, Farfel J M, Ferretti R E, Leite R E *et al* 2009 Equal numbers of neuronal and non-neuronal cells make the human brain an isometrically scaled-up primate brain *J. Comp. Neurol.* **513** 532–41

Baddeley R, Abbott L F, Booth M C, Sengpiel F, Freeman T, Wakeman E A and Rolls E T 1997 Responses of neurons in primary and inferior temporal visual cortices to natural scenes *Proc. Biol. Sci.* **264** 1775–83

Bear G 1998 *Slant* (New York: Tor)

Benet J 2017 Filecoin: a decentralized storage network. *Protocol Labs* https://ipld.io/ (accessed 6 May 2020)

Blaschka T 2020 How the world's largest banks use advanced graph analytics to fight fraud *TigerGraph* https://tigergraph.com/2020/01/23/how-the-worlds-largest-banks-use-advanced-graph-analytics-to-fight-fraud/ (accessed 6 May 2020)

Bozinovski S, Sestakov M and Bozinovska L 1988 Using EEG alpha rhythm to control a mobile robot *Proc. IEEE Annual Conf. of Medical and Biological Society (New Orleans)* G Harris and C Walker 1515–16

Branca M A 2005 Gene therapy: cursed or inching towards credibility? *Nat. Biotechnol.* **23** 519–21

Brookes N J and Locatelli G 2015 Power plants as megaprojects: using empirics to shape policy, planning, and construction management *Util. Policy* **36** 57–66

Carminati F 2018 Quantum thinking required *CERN Cour.* **58** 5 https://cern-courier.web.cern.ch/a/viewpoint-quantum-thinking-required/ (accessed 3 October 2025)

Cavalcanti A, Hogg T, Shirinzadeh B and Liaw H C 2006 Nanorobot communication techniques: a comprehensive tutorial *IEEE ICARCV 2006 Int. Conf. on Control, Automation, Robotics and Vision* 1–6

Chao S, Roberts J S, Marteau T M, Silliman R, Cupples L A and Green R C 2008 Health behavior changes after genetic risk assessment for Alzheimer disease: the REVEAL study *Alzheimer Dis. Assoc. Disord.* **22** 94–7

Chiang T 2002 *Arrival* (New York: Vintage Books)

Chudler E H 2009 Brain facts and figures *Neuroscience for Kids* http://faculty.washington.edu/chudler/facts.html (accessed 6 May 2020)

Cisco 2020 Cisco annual internet report, 2018–2023 *Cisco White paper* https://www.cisco.com/c/en/us/solutions/collateral/executive-perspectives/annual-internet-report/white-paper-c11-741490.html (accessed 3 October 2025)

Cranshaw J and Kittur A 2011 The polymath project: lessons from a successful online collaboration in mathematics *CHI '11: Proc. of the SIGCHI Conf. on Human Factors in Computing Systems* (New York, NY: ACM) 1865–74

Domschke A and Boehm F J 2017 Application of a conceptual nanomedical platform to facilitate the mapping of the human brain: survey of cognitive functions and implications *The Physics of the Mind and Brain Disorders Integrated Neural Circuits Supporting the Emergence of Mind* ed I Opris and M F Casanova (New York, NY: Springer)

Ehrsam F 2017 Scaling ethereum to billions of users *Medium* https://medium.com/@FEhrsam/scaling-ethereum-to-billions-of-users-f37d9f487db1 (accessed 6 May 2020)

Fahy G M and Wowk B 2015 Principles of cryopreservation by vitrification *Methods Mol. Biol.* **1257** 21–82

Finger S 1994 *Origins of Neuroscience: A history of explorations into brain function* (Oxford: Oxford University Press)

Feynman R P 1985 Quantum mechanical computers *Found. Phys.* **16** 507–31

Feynman R P, Leighton R B and Sands M 2005 *The Feynman Lectures on Physics: The Definitive and Extended Edition* 2nd edn (New York: Addison Wesley)

Flyvbjerg B 2017 *The Oxford Handbook of Megaproject Management* (Oxford: Oxford University Press)

Freitas R A Jr. 2000 Clottocytes: artificial mechanical platelets *Foresight Update* **41** 9–11 http://imm.org/Reports/Rep018.html (accessed 6 May 2020)

Freitas R A Jr. 2005 Microbivores: artificial mechanical phagocytes using digest and discharge protocol *Journal of Evolution and Technology* **14** 55–106 http://jetpress.org/volume14/freitas.pdf (accessed 6 May 2020)

Freitas R A Jr. 2012 Welcome to the future of medicine *The Transhumanist Reader* ed M More and N Vita-More (Oxford: Wiley-Blackwell) 67–72

Freitas R A Jr. and Phoenix C J 2002 Vasculoid: a personal nanomedical appliance to replace human blood *Journal of Evolution and Technology* **11** 1–139 http://jetpress.org/volume11/vasculoid.html (accessed 6 May 2020)

Gerstner W, Kreiter A K, Markram H and Herz A V M 1997 Neural codes: firing rates and beyond *PNAS* **94** 12740–41

Gibson W 1984 *Neuromancer* (New York: Ace (Random House))

Gonzalez-Raya T, Solano E and Sanz M 2020 Quantized three-ion-channel neuron model for neural action potentials arXiv:1906.07570v2 [q-bio.NC]

Green R C, Roberts J S, Cupples L A, Relkin N R, Whitehouse P J, Brown T *et al* 2009 REVEAL study group. Disclosure of APOE genotype for risk of Alzheimer's disease *N. Engl. J. Med.* **361** 245–54

Grumbling E and Horowitz M 2019 *Quantum Computing: Progress and Prospects* (Washington DC: U.S. National Academies of Sciences)

Harms M 2016 *Crystal Society* (Creative Commons Attribution-NonCommercial 4.0 International license) v1.1.9 http://crystal.raelifin.com (accessed 6 May 2020)

Harris S A and Kendon V M 2010 Quantum-assisted biomolecular modelling *Phil. Trans. R. Soc. A* **368** 3581–92

Hartnoll S A, Lucas A and Sachdev S 2018 *Holographic Quantum Matter* (Cambridge MA: MIT Press)

Hashimoto K, Sugishita S, Tanaka A and Tomiya A 2018 Deep learning and the AdS/CFT correspondence *Phys. Rev. D.* **98** 046019

Hogg T and Freitas R A Jr. 2009 Chemical power for microscopic robots in capillaries *Nanomed. Nanotechnol. Biol. Med.* **6** 298–317

Kaku M 2018 *The Future of Humanity* (New York: Doubleday)

Kalinin K P and Berloff N G 2018 Blockchain platform with proof-of-work based on analog Hamiltonian optimisers arXiv:1802.10091 [quant-ph]

Kardashev N 1964 Transmission of information by extraterrestrial civilizations *Sov. Astron.* **8** 217–21

Kashtan M 2014 *Reweaving Our Human Fabric: Working Together to Create a Nonviolent Future* (Oakland CA: Fearless Heart Publications)

Keenan T 2017 CloudMinds, the world's first cloud robot operator, launches mobile-intranet cloud services, enabling secure cloud robotic deployments, and data A1 handset, world's first mobile phone robotic control unit *Business Wire* https://businesswire.com/news/home/20170912006534/en/CloudMinds-World%E2%80%99s-Cloud-Robot-Operator-Launches-Mobile-Intranet (accessed 11 May 2024)

Kendon V M, Nemoto K and Munro W J 2010 Quantum analogue computing *Phil. Trans. R. Soc.* **368** 3609–20

Kettenmann H and Verkhratsky A 2011 Neuroglia—living nerve glue *Fortschr. Neurol. Psychiatr.* **79** 588–97

Kleinfeld D, Bharioke A, Blinder P, Bock D D, Briggman K L, Chklovskii D B *et al* 2011 Large-scale automated histology in the pursuit of connectomes *J. Neurosci.* **31** 16125–38

Kraft C C Jr. 2001 *Flight: My Life in Mission Control* (New York: Dutton Penguin Group)

Kress N 1994 *Beggars and Choosers* (New York: Tor)

Kuffler S W and Yoshikami D 1975 The number of transmitter molecules in a quantum: an estimate from iontophoretic application of acetylcholine at the neuromuscular synapse *J. Physiol.* **251** 465–82

LeCun Y, Bengio Y and Hinton G 2015 Deep learning *Nature* **521** 436–44

Ley B 1999 Diameter of a human hair *The Physics Factbook* https://hypertextbook.com/facts/1999/BrianLey.shtml (accessed 6 May 2020)

Maldacena J 1999 The large N limit of superconformal field theories and supergravity *Int. J. Theor. Phys.* **38** 1113–33

Maldacena J 2012 The gauge/gravity duality *Black Holes in Higher Dimensions* ed G T Horowitz (Cambridge: Cambridge University Press) 325–47

Marosfoi M G, Korin N, Gounis M J, Uzun O, Vedantham S, Langan E T *et al* 2015 Shear-activated nanoparticle aggregates combined with temporary endovascular bypass to treat large vessel occlusion *Stroke* **46** 3507–13

Martins N R B *et al* 2019 Human brain/cloud interface *Front. Neurosci.* **13** 1–24

Martins N R B, Erlhagen W and Freitas R A Jr. 2016 Human connectome mapping and monitoring using neuronanorobots *J. Evol. Technol.* **26** 1–24 https://jetpress.org/v26.1/martins.htm

Martins N R B, Erlhagen W and Freitas R A Jr. 2012 Non-destructive whole-brain monitoring using nanorobots: neural electrical data rate requirements. *Int. J. Mach. Conscious.* **4** 109–40

McLeod S 2007 Maslow's hierarchy of needs *Simply Psychology* https://simplypsychology.org/maslow.html (accessed 6 May 2020)

Meyer D, Bonhoeffer T and Scheuss V 2014 Balance and stability of synaptic structures during synaptic plasticity *Neuron* **82** 430–43

Mitchell N, Petralia R S, Currier D G, Wang Y X, Kim A, Mattson M P and Yao P J 2012 Sonic hedgehog regulates presynaptic terminal size, ultrastructure and function in hippocampal neurons *J. Cell Sci.* **125** 4207–13

Monroe D 2014 Neuromorphic computing gets ready for the (really) big time *Commun. ACM* **57** 13–5

Morris R and Fillenz M 2003 *Neuroscience: The Science of the Brain* (Liverpool: The British Neuroscience Association)

Musk E 2019 An integrated brain–machine interface platform with thousands of channels *J. Med. Int. Res.* **21** e16194

Nave K 2016 How Craig Venter is fighting ageing with genome sequencing *Wired UK* https://wired.co.uk/article/craig-venter-human-longevity-genome-diseases-ageing (accessed 6 May 2020)

Nelson P C 2008 *Biological Physics: Energy, Information, Life* (New York: W.H. Freeman & Co Ltd.)

Ng A 2025 Machine Learning [Online course]. Coursera. https://www.coursera.org/learn/machine-learning (accessed October 20, 2025)

Nicholls J G, Martin R A, Brown D A, Diamond M E, Weisblat D A and Fuchs P A 2012 *From Neuron to Brain* 5th edn (Sunderland MA: Sinauer Associates, Inc.)

Nicolas-Alonso L F and Gomez-Gil J 2012 Brain computer interfaces, a review *Sensors (Basel)* **12** 1211–79

NIH (National Institute on Deafness and Other Communication Disorders) 2011 *NIH Publication No. 11–4798. Cochlear Implants* http://nidcd.nih.gov/health/hearing/pages/coch.aspx (accessed 6 May 2020)

Olson J, Cao J, Romero P, Johnson P-L, Dallaire-Demers N, Sawaya P *et al* 2016 Quantum information and computation for chemistry *NSF Workshop Report* National Science Foundation https://arxiv.org/abs/1706.05413

Parent A 1996 *Carpenter's Human Neuroanatomy* 9th edn (London: Williams & Wilkins)

Polymath D H J 2010 Density Hales-Jewett and Moser numbers *Bolyai Society Mathematical Studies* ed I Bárány, J Solymosi and G Sági (Berlin: Springer) **21** 689–753

Polymath D H J 2012 A new proof of the density Hales-Jewett theorem *Ann. Math.* **175** 1283–327

Poirazi P, Brannon T and Mel B W 2003 Pyramidal neuron as two-layer neural network *Neuron* **37** 989–99

Poon J and Dryja T 2016 Lightning network paper, v0.5.9.1 https://cryptochainuni.com/wp-content/uploads/Bitcoin-lightning-network-paper-DRAFT-0.5.pdf (accessed 6 May 2020)

Procopio L M, Moqanaki A, Araújo M, Costa F, Alonso Calafell I, Dowd E G *et al* 2015 Experimental superposition of orders of quantum gates *Nat. Commun.* **6** 1–6

Proust M 1929 The captive *Search of Lost Time* vol V ed C K Scott Moncrieff (New York: Random House)

Raymond E S 1999 *The Cathedral and the Bazaar* (Sebastopol CA: O'Reilly Media)

Sapra N V, Yang K Y, Vercruysse D, Leedle K J, Black D S, England R J *et al* 2020 On-chip integrated laser-driven particle accelerator *Science* **367** 79–83

Scimemi A and Beato M 2009 Determining the neurotransmitter concentration profile at active synapses *Mol. Neurobiol.* **40** 289–306

Scroxton A 2018 Medical firm debuts internet-connected prosthetic limbs *Comput. Wkly.* https://www.computerweekly.com/news/252448904/Medical-firm-debuts-internet-connected-prosthetic-limbs

Sergio A and Pires T 2014 *AdS/CFT Correspondence in Condensed Matter* (San Rafael, CA: IOP Concise Physics Morgan & Claypool Publishers)

Shepherd G M 1974 *The Synaptic Organization of the Brain. An Introduction.* (New York: Oxford University Press)

Shih J J, Krusienski D J and Wolpawc J R 2012 Brain–computer interfaces in medicine *Mayo. Clin. Proc.* **87** 268–79

Singh S, Lu S, Kokar M M and Kogut P A 2017 Detection and classification of emergent behaviors using multi-agent simulation framework (WIP) *IEEE. Spring 2017 Simulation Multi-Conf. Society for Modeling & Simulation (SCS)* 4650–51 https://www.academia.edu/66100173/Detection_and_Classification_of_Emergent_Behaviors_using_Multi_Agent_Simulation_Framework_WIP_

Skinner C 2017 The largest payments company in the world most people have never heard of *The Next Web* https://thenextweb.com/asia/2017/03/06/largest-payments-company-world-people-never-heard/ (accessed 6 May 2020)

Strohmeyer R 2008 The 7 worst tech predictions of all time *PCWorld* https://pcworld.com/article/155984/worst_tech_predictions.html (accessed 6 May 2020)

Stross C 2013 *Neptune's Children* (New York: Penguin)

Susskind L 1995 The world as a hologram *J. Math. Phys.* **36** 6377–96

Swan M 2015a *Blockchain: Blueprint for a New Economy* (Sebastopol CA: O'Reilly Media)

Swan M 2015b Blockchain thinking: the brain as a DAC (decentralized autonomous corporation) *Technol. Soc. Mag.* **34** 41–52

Swan M 2015c Digital simondon: the collective individuation of man and machine Gilbert Simondon: Media and technics. Special issue *Platform: J. Med. Commun.* **6** 46–58

Swan M 2016 The future of brain–computer interfaces: blockchaining your way into a cloudmind *J. Evol. Technol.* **26** 60–81

Swan M 2019 Transhuman crypto cloudminds *The Transhuman Handbook* ed N Lee (Switzerland: Springer) 513–27

Swan M 2020 Black hole zero-knowledge proofs *FQXi Essay Contest* (Undecidability, Uncomputability, and Unpredictability) https://fqxi.org/community/forum/topic/3442 (accessed 6 May 2020)

Swan M 2018 Technophysics, smart health networks, and the bio-cryptoeconomy: quantized fungible global health care equivalency units for health and wellbeing ed F Boehm *Nanotechnology, Nanomedicine, and AI: Toward the Dream of Global Health Care Equivalency* (Boca Raton FL: CRC Press) (forthcoming)

Swan M, dos Santos R P, Lebedev M A and Witte F 2022 *Quantum Computing for the Brain* (London: World Scientific)

Swan M, dos Santos R P and Witte F 2020 *Quantum Computing: Physics, Blockchains, and Deep Learning Smart Networks* (London: World Scientific)

Tuckman B W 1965 Developmental sequence in small groups *Psychol. Bull.* **63** 384–99

van Albada S J, Rowley A G and Senk J 2018 Performance comparison of the digital neuromorphic hardware SpiNNaker and the neural network simulation software NEST for a full-scale cortical microcircuit model *Front. Neurosci.* **12** 291

von Bartheld C S, Bahney J and Herculano-Houzel S 2016 The search for true numbers of neurons and glial cells in the human brain: a review of 150 years of cell counting *J. Comp. Neurol.* **524** 3865–95

Vinge V 2007 *Rainbow's End* (New York: Tor)

Visa 2011 Visa transactions hit peak on dec. 23, 2011 *Visa Viewpoints* https://visa.com/blogarchives/us/2011/01/12/visa-transactions-hit-peak-on-dec-23/index.html (accessed 6 May 2020)

Wehner S, Elkouss D and Hanson R 2018 Quantum internet: a vision for the road ahead *Science* **362** eaam9288

Williams S 2017 Optogenetic therapies move closer to clinical use *TheScientist* https://www.the-scientist.com/optogenetic-therapies-move-closer-to-clinical-use-30611

Yaffe P 2011 The 7% rule: fact, fiction, or misunderstanding *Ubiquity* **10** 1–5

Yong E 2016 *I Contain Multitudes: The Microbes within Us and Grander View of Life* (New York: Ecco Press)

Wang Q, Ding S L, Li Y, Royall J, Feng D, Lesnar P *et al* 2020 The Allen mouse brain common coordinate framework: a 3D reference atlas *Cell* **181** 1–18

Zao J K, Gan T T, You C K, Chung C E, Wang Y T, Rodríguez Méndez S J *et al* 2014 Pervasive brain monitoring and data sharing based on multi-tier distributed computing and linked data technology *Front. Human Neurosci.* **8** 1–16

IOP Publishing

Nanomedical Brain/Cloud Interface
Explorations and implications
Frank J Boehm

Chapter 6

The ultimate chip

Howard Bloom

Howard Bloom, the author of eight books, has been called the Einstein, Newton, Darwin and Freud of the 21st century by Britain's Channel4 TV. He has published in journals or given lectures at scholarly conferences in twelve different scientific fields, from quantum physics and cosmology to evolutionary science, information science, and aerospace.

The age of intelligent AI agents is upon us. They appear out of the blue one day on our smartphones, cocooned within the latest innocuous software updates, which some wary technophobes might perceive as Trojan horses. Now they listen, cheerily strike up conversations, and eagerly engage and encourage our every move, no matter how bizarre, while satisfying virtually every whim. Serving as digitized Aladdin's lamps, they are activated by our queries, which are met with almost instantaneous responses, helping us navigate our increasingly complex worlds. That first nascent AI agent in ~1966, manifested as the rule-based ELIZA system, has evolved over the last four decades to emerge as the autonomous adaptive large language model (LLM)-driven AI agents that dwell among us today. The exponential advances witnessed in synergies between LLMs and distributed agentic models have translated into potent capacities for intricate cognition, contextualization, and automatic decision-making, all put to good use toward knowing what we desire before we do. A brain–cloud interface could take this many steps further, when our highly personalized trusted digital confidants are at our beck and call—summoned by merely thinking of them, ready to advise, act, or create.

And the man who foresaw it all was a scientific thinker and the author of eight books, Howard Bloom.

6.1 Aladdin's lamp

Once upon a time, the idea of a genie in Aladdin's lamp was a fantasy. That genie was conjured in prose sometime before 1704 in the Arabian Nights (Haddawy 2008).

doi:10.1088/978-0-7503-2144-0ch6   6-1   © IOP Publishing Ltd 2025. All rights,
including for text and data mining (TDM), artificial intelligence (AI) training, and similar technologies, are reserved.

But it could not be conjured in reality. Now, with the coming interface between the brain and the cloud (Martins *et al* 2019), that fantasy is about to come true. Yes, within a decade, we may well have the genie of Aladdin's dreams, the device that will read your thoughts, figure out your needs before you know you have them, and fetch you the perfect girlfriend, the perfect shoulder to cry on, the perfect book, the perfect doctor, or the perfect tool.

Meanwhile, once upon a time, from roughly 1620 until this very minute, advertising was a tremendous waste of money, energy, and time. To reach one person who needed your product, you had to spread your message to millions who did not need it. And you had to repeat your exposures until your name stuck. For every sale, you had to hammer millions of people who did not need your offering. In fact, you had to hammer those millions over and over again. Now, we are on the verge of reversing that and allowing a customer to reach out to you directly when your gizmo or service is exactly what he or she needs. This is all coming thanks to the ultimate chip and something called an intelligent agent.

6.2 Boolean algebra and the ultimate chip

Here is how I got into this territory: I built my first Boolean algebra machine (Knuth 2003) and codesigned my first computer when I was twelve. The computer won some science fair awards. Twelve years later, I studied early neuroscience with E E Coons, the man who discovered what the hypothalamus did (Coons *et al* 1965).

I would eventually become a visiting scholar at NYU's graduate psychology department and would publish in journals or give lectures at scholarly conferences in twelve different scientific fields, from quantum physics and cosmology to evolutionary biology, information science, and aerospace.

But around 1970, I wrote a science fiction story. In it, a man who has racked up hundreds of unpaid traffic tickets goes before a judge. The judge gives him two choices—six months in jail or participation in a scientific study. The traffic violator chooses the scientific study. So, he is implanted with a chip. A chip that monitors his thoughts and his emotions. And he is sent an inflatable woman, a sophisticated sex toy. The experiment involves measuring his responses to his inflatable female. Naturally, he falls in love with the sex toy. Then tragedy strikes. With an accidental swipe of a broken toenail, our hero punctures his artificial her, lets the air out, and loses her forever. The relevant part of the story is not our protagonist's tragic loss. It is the chip. And the idea of chips like this implanted in hundreds of thousands of people so that scientists can study emotion, behavior, and the mysteries of the psyche and of the crowd in a whole new way. The harvest of insight could be huge. And privacy problems never occurred to me. That is how I started to conceptualize something I called the ultimate chip.

6.3 Encephalomyelitis/chronic fatigue syndrome

Then, in the 1990s, I was stuck in bed with a ferocious illness—myalgic encephalomyelitis (ME), or chronic fatigue syndrome (CFS) (Grach *et al* 2023). For five years, I was too weak to speak and too weak to have another person in the room with me. As all the things I had imagined for my future became impossible, I lost my sense of humanity. The only real estate in which I could travel, make new friends, and build a new identity was the internet. I was already a technology nut. But this experience taught me about the intense emotional possibilities of the cloud.

After five years, I regained the strength to use my voice. I was still stuck in my bedroom, but I could have visitors. A 30-year-old computer scientist from Boston started making pilgrimages from Massachusetts to New York to see me. Why me? My visitor had put together a set of trading cards with the world's geniuses, people like Albert Einstein, Richard Feynman, Isaac Newton, and, of all people, me. The genius trading card was conceived by Alexander Chislenko. He was from Russia and had moved in 1969 from Leningrad to Boston to be close to computer pioneers such as Marvin Minsky (Minsky 1954). And Alex had bet his career on a field that had become radically unfashionable, something called 'artificial intelligence' (Chislenko 2024).

Yes, artificial intelligence had been popular in the 1980s. Then, in the 1990s, scientific fashion tossed it aside as an unproductive distraction. Yet Chislenko hung in there. He had a simple concept, simple but brilliant.

6.4 Intelligent agents

In your computer and cellphone, you would have an intelligent agent (Roose 2023). A piece of software that would belong to you, that would learn your needs, your excitements, your irritations, your pains, your desires, and your dreams. A piece of software that would come to know you better every day. That intelligent agent would do something remarkable for you. It would surf the cloud 24/7 bargaining with the intelligent agents of the world's other eight billion people, snooping, hunting, poking, and probing. For what? For other intelligent agents offering things that fit your needs. Or, to put it differently, checking other intelligent agents to see if their needs fit yours. And to offer the possibility of connecting you with the owner of another intelligent agent if you choose to follow up on your intelligent agent's suggestion.

Today, the primitive ancestors of these intelligent agents are alive in the product suggestion algorithms (Alsobhi and Amare 2022) of Netflix and Amazon. But they belong to Netflix and Amazon. Not to you. And marketers are working madly to use deep data to see your needs and deliver the right offering to you at precisely the right moment, just when you need it most. But there are extreme concerns about invasions of your privacy. Why? Because these harvests of data are out to read your mind. But they belong to companies, some of whom are trying to prey upon you, not serve you. And, most importantly, because these harvests of data about you do not belong to you.

In Alexander Chislenko's vision, these harvests of data about you would belong to you. Exclusively. Which means turning advertising on its head. To the benefit of both you and the advertiser. Your intelligent agent and those of billions of others would come to the advertiser when you needed her goods or services the most. Your intelligent agents and those of billions of others would eliminate the need for advertising to hundreds of millions who do not need a company's offerings.

6.5 Golden age of radio

This is the opposite of a revolution that occurred in the early days of radio. In 1895, the inventors of radio were looking for a way to send telegrams without using wires. When they finally came up with a way to use radio waves, they had a problem. Telegraph lines go from one point to another. Radio waves do not. They spread out in circles. This was a huge disadvantage if you wanted to send a telegram from your home in San Francisco to your aunt Maud in London. Then someone came up with a way to turn this problem into an opportunity, to turn a disadvantage into a blessing. The new concept that turned radio's flaw into a gold mine was called 'broadcasting' (Balogh 2023). And it led to the golden age of radio, to the big three TV networks, to celebrities from Arthur Godfrey to Jimmy Kimmel, and to the nightly news.

6.6 Personalized Ashley

Alexander Chislenko's intelligent agent, the agent that belongs to you and to you alone, the agent that works tirelessly to learn you so well that it can read your mind, takes advertising from a broadcast medium back to point-to-point communication.

This intelligent agent could lodge in a chip behind your right ear and could do amazing things for you. Say you are on a business trip to Detroit. You check into your hotel, drop your bags in your room, and go down to the bar hoping for human company. Sure enough, there are other folks in there. But you do not know them. They have off-putting expressions on their faces. So you buy a drink, nurse it, then go back up to your room without having met a soul. Your chip and its intelligent agent would change all that. As you scanned the room, it would use facial recognition and access to the cloud to fetch the names of the folks you had looked forlornly at, get their biographies, and pick the ones with the interests and personalities that best fit yours. Your intelligent agent lodged in the ultimate chip would hand you the results and suggest opening questions with which you could approach each one—like 'Hi, I hear you're into technology. So am I.'

Then there are those mornings when you wake up, stumble to the bathroom, and are hit with a brilliant idea. The idea is so amazing, you know you will remember it long enough to write it down. But you have more urgent things to attend to. When you are finished and walk back to your room, the idea is gone. As if it never existed. Those days are over. Your chip will store your brilliant idea in the cloud. Along with all of your other spectacular brain flashes. And when you are in the middle of a project or a conversation that those old thoughts might apply to, the intelligent agent in your chip will remind you of your relevant past astonishing ideas and ask if you want them.

But that is not all. Those speeches that you write but do not have the time or patience to memorize? The intelligent agent in your chip will recall them and feed them to you word after word when the time comes to deliver them to an audience. You will never be tongue-tied when speaking in public again.

And those miserable days when your boss comes down on you like a ton of bricks? You walk out of the office at the end of the day wounded. But you have learned that if you tell your wife about your misery, she, too, will attack you. Yes, that is a nasty pattern that even shows up in seagull behavior—attacking your fellow creatures in pain, the ones who need you most. So who can you talk to? The one being that spends more time caring about you than anyone else in the world. Your intelligent agent. Let us call her Ashley. When you get into your car, you pour your heart out to her. She listens sympathetically. And she scouts the web to see if there is anything that might help ease your agony. Like a job at a company that would appreciate you.

What tools will make your genie in a bottle—your intelligent agent—come to life? Your smartphone camera, Paul Ekman's microexpressions (Kemeny *et al* 2012, Guerdelli *et al* 2022, Saffaryazdi *et al* 2022), facial recognition technology (Kawaguchi *et al* 2024, Guo *et al* 2025, Kim *et al* 2025), prosody-reading sound technologies (Pethe *et al* 2023, Kane *et al* 2024, Wang *et al* 2024) and your smartwatch sensing your heart rate, blood pressure, and galvanic skin response (GSR) (Bennett *et al* 2022). Not to mention the very technology that was shunned for a decade as a nonstarter, a loser: AI, artificial intelligence.

6.7 Neuralink

Then there is the chip Elon Musk is developing. At a time when some neuroscience labs are still using single-electrode probes in their research, the device from Musk's Neuralink company, whose first-generation N1 chip had 1024 electrodes, will have 16 000 electrodes in next-generation models (Hitti 2019, Musk 2019, Shankland 2022, Drew 2024a, 2024b, Krämer 2024). The short-term goal is to implant Musk's chip in the brains of quadriplegics so those unfortunates can regain some of their abilities. But the long-term goal is to enhance the powers of you and me.

6.8 The ultimate chip

When I first started figuring out the ultimate brain–cloud interface in 1970 and began the project I called the ultimate chip, computers were massive and expensive. So, the ultimate chip was impossible. Each of the desktop computers we used in Fortran courses at NYU, where I went to school, covered the surface of a large table. Each computer was four feet wide, three feet deep, and a foot and a half tall. We had six of these big, flat devices in one room. The machines were colored gray, but they were black boxes.

The idea of hooking them to a monitor and a keyboard had not yet caught on (Bellis 2006). Much less the idea of connecting them to your brain. You communicated with these 'desktops' via punched cards. In other words, you wrote a program, converted it to a two-inch-high stack of punched cards, and fed the 200 punched cards one by one into the computer the way you feed your credit card into a credit card reader today. Then the Radio Shack folks and Toshiba invented the first laptop computers. And by roughly 2000, those computers had microphones and video cameras as standard equipment. In other words, by roughly 2000, computers had the ability to watch and listen to you (Minton *et al* 2014).

Meanwhile, way back in roughly 1999, University of San Francisco psychologist Paul Ekman issued pictures of 110 microexpressions, quick flashes of facial expression that revealed your emotions (Donato *et al* 1999). Expressions that a computer equipped with a webcam could someday use to gauge your inner life. Then came the iPhone in 2007 and put the power to listen to you all day long in your pocket. And the power to serve you. In 2010, Demis Hassabis founded DeepMind Technologies and popularized the concept of deep learning (De Fauw *et al* 2018). A form of learning that could eventually learn the innermost you. In roughly 2010, computers became good at facial recognition (Hsieh *et al* 2010). And in 2017, the Israeli company Beyond Verbal introduced an application programming interface (API) that could read the emotions in your voice—a program that could read your prosody, the music of your speech, not just the words (Tanner 2021).

Finally, in 2014 came Apple's smartwatch with sensors for your heart rate and your steps (Golbus *et al* 2021). Other wearables emerged with the ability to sense your blood pressure (Arakawa 2018, Kario 2020, Ismail *et al* 2023, Min *et al* 2025). These were all the tools an intelligent agent would need to read your mind and your feelings. All the tools the ultimate chip would need to learn to anticipate your needs. They are all the tools that could give you a brain–cloud interface. If the intelligent agent and the ultimate chip ever come to be, Amazon and other online services will gladly provide the goods and services that will help you achieve your goals.

But the most important needs in your life are not goods and services. They are human contacts. They are love and warmth. And your intelligent agent will be capable of connecting you with people you should meet and with people of the sex you prefer who might be ripe for a relationship. But, alas, today computer mindreading is being designed to be top-down. Not bottom-up. Alexander Chislenko's personal, private intelligent agents have not arrived. Yet.

In 2010, I was on a panel in San Diego at Demo—an event where next-tech developers pitch their products to venture capitalists. On the panel with me was Peter Norvig, director of research for Google. I preached the genie-in-the-bottle intelligent agent to the audience. Peter preached a statistical exploration of massive amounts of data—all the data of Google plus all the data of YouTube and all the data of Google image search. Not to mention the data of Gmail. Peter preached top-down exploration of deep data, very deep indeed. And very top down. To date, Peter's vision has won out.

But there is a lot more development of the ultimate chip to come. Will computers ever go past implantable chips and be able to read your mind without insisting that you wear special hardware? In other words, could the ultimate chip someday work without a chip? I suspect that capability is not far away.

But, more important, will Alexander Chislenko's vision, the personal intelligent agent vision, ever come to pass? Will the advertising model ever turn upside down from broadcasting to narrowcasting? From messages sent by the advertiser to messages sent by you? Will the artificial intelligence that reads your mind ever belong to you? We shall have to see. But one way or the other, the brain–cloud interface is on its way to achieving a simple goal: becoming your genie in a bottle. Serving your every need.

References

Alsobhi A and Amare N 2022 Ontology-based relational product recommendation system *Comput. Math. Methods Med.* **2022** 1591044

Arakawa T 2018 Recent research and developing trends of wearable sensors for detecting blood pressure *Sensors (Basel)* **18** 2772

Balogh D 2023 *Telecommunications History: Broadcasting* (Ooma, Inc.) https://ooma.ca/blog/telecommunications-history-broadcasting/ (accessed 16 May 2024)

Bellis M 2006 *History of the Computer Keyboard* (About, Inc.) https://theinventors.org/library/inventors/blcomputer_keyboard.htm (accessed 19 May 2024)

Bennett J P, Liu Y E, Kelly N N, Quon B K, Wong M C, McCarthy C *et al* 2022 Next-generation smart watches to estimate whole-body composition using bioimpedance analysis: accuracy and precision in a diverse, multiethnic sample *Am. J. Clin. Nutr.* **116** 1418–29

Chislenko A 2024 Great thinkers and visionaries. http://lucifer.com/~sasha/thinkers.html (accessed 13 May 2024)

Coons E E, Levak M and Miller N E 1965 Lateral hypothalamus: learning of food-seeking response motivated by electrical stimulation *Science* **150** 1320–1

De Fauw J, Ledsam J R, Romera-Paredes B, Nikolov S, Tomasev N, Blackwell S *et al* 2018 Clinically applicable deep learning for diagnosis and referral in retinal disease *Nat. Med.* **24** 1342–50

Donato G, Bartlett M S, Hager J C, Ekman P and Sejnowski T J 1999 Classifying facial actions *IEEE Trans. Pattern Anal. Mach. Intell.* **21** 974

Drew L 2024a Elon Musk's Neuralink brain chip: what scientists think of first human trial *Nature*

Drew L 2024b Neuralink brain chip: advance sparks safety and secrecy concerns *Nature* **627** 19

Golbus J R, Pescatore N A, Nallamothu B K, Shah N and Kheterpal S 2021 Wearable device signals and home blood pressure data across age, sex, race, ethnicity, and clinical phenotypes in the Michigan Predictive Activity & Clinical Trajectories in Health (MIPACT) study: a prospective, community-based observational study *Lancet. Digit. Health.* **3** e707–15

Grach S L, Seltzer J, Chon T Y and Ganesh R 2023 Diagnosis and management of myalgic encephalomyelitis/chronic fatigue syndrome *Mayo Clin. Proc.* **98** 1544–51

Guerdelli H, Ferrari C, Barhoumi W, Ghazouani H and Berretti S 2022 Macro-and micro-expressions facial datasets: a survey *Sensors (Basel)* **22** 1524

Guo Y, Zhai P, Jia D and Li W 2025 A cryptosystem for face recognition based on optical interference and phase truncation theory *Sci. Rep.* **15** 25371

Haddawy H 2008 *The Arabian Nights = Alf Laylah wa-Laylah* (New York: W. W. Norton & Co Inc)

Hitti N 2019 Elon Musk's Neuralink implant will 'merge' humans with AI *dezeen* https://dezeen.com/2019/07/22/elon-musk-neuralink-implant-ai-technology/ (accessed 19 May 2024)

Hsieh C K, Lai S H and Chen Y C 2010 An optical flow-based approach to robust face recognition under expression variations *IEEE Trans. Image Process.* **19** 233–40

Ismail S N A, Nayan N A, Mohammad Haniff M A S, Jaafar R and May Z 2023 Wearable two-dimensional nanomaterial-based flexible sensors for blood pressure monitoring: a review *Nanomaterials (Basel)* **13** 852

Kane J, Johnstone M N and Szewczyk P 2024 Voice synthesis improvement by machine learning of natural prosody *Sensors (Basel)* **24** 1624

Kario K 2020 Management of hypertension in the digital era: small wearable monitoring devices for remote blood pressure monitoring *Hypertension* **76** 640–50

Kawaguchi T, Ono K and Hikawa H 2024 Electroencephalogram-based facial gesture recognition using self-organizing map *Sensors (Basel)* **24** 2741

Kemeny M E, Foltz C, Cavanagh J F, Cullen M, Giese-Davis J, Jennings P *et al* 2012 Contemplative/emotion training reduces negative emotional behavior and promotes prosocial responses *Emotion* **12** 338–50

Kim J H, Cha H S and Im C H 2025 Facial electromyogram-based emotion recognition for virtual reality applications using machine learning classifiers trained on posed expressions *Biomed. Eng. Lett.* **15** 773–83

Knuth K H 2003 Intelligent machines in the twenty-first century: foundations of inference and inquiry *Philos Trans. A. Math. Phys. Eng. Sci.* **361** 2859–73

Krämer K 2024 AI & robotics briefing: lack of transparency surrounds Neuralink's 'brain-reading' chip *Nature*

Martins N R B *et al* 2019 Human brain/cloud interface *Front. Neurosci.* **13** 112

Min S, An J, Lee J H, Kim J H, Joe D J, Eom S H *et al* 2025 Wearable blood pressure sensors for cardiovascular monitoring and machine learning algorithms for blood pressure estimation *Nat. Rev. Cardiol* **22** 629–48

Minsky M 1954 *Theory of Neural-Analog Reinforcement Systems and its Application to the Brain-Model Problem* (Princeton UniversityProQuest Dissertations Publishing) 0009438

Minton S, Allan M and Valdes W 2014 Teleneonatology: a major tool for the future *Pediatr. Ann.* **43** e50–5

Musk E 2019 Neuralink. An integrated brain–machine interface platform with thousands of channels *J. Med. Internet Res.* **21** e16194

Pethe C, Pham B, Childress F D, Yin Y and Skiena S 2023 Prosody analysis of audiobooks *arXiv preprint* arXiv:2310.06930

Roose K 2023 Personalized A.I. agents are here. Is the world ready for them? *The New York Times* https://nytimes.com/2023/11/10/technology/personalized-ai-agents.html (accessed 16 May 2024)

Saffaryazdi N, Wasim S T, Dileep K, Nia A F, Nanayakkara S, Broadbent E and Billinghurst M 2022 Using facial micro-expressions in combination with EEG and physiological signals for emotion recognition *Front. Psychol.* **13** 864047

Shankland S 2022 Neuralink's upgraded brain chip hopes to help the blind see and the paralyzed walk *CNET* https://cnet.com/science/neuralink-upgraded-brain-chip-hopes-to-help-the-blind-see-and-the-paralyzed-walk/ (accessed 17 July 2025)

Tanner A 2021 Can technology read your emotions? *Consumer Reports* (Inc.) https://consumer-reports.org/artificial-intelligence/can-technology-read-your-emotions-a1096874808/ (accessed 19 May 2024)

Wang K, Qiao X, Sammit G, Larson E C, Nese J and Kamata A 2024 Improving automated scoring of prosody in oral reading fluency using deep learning algorithm *Front. Educ.* **9** 1440760

www.ingramcontent.com/pod-product-compliance
Lightning Source LLC
Chambersburg PA
CBHW080549220326
41599CB00032B/6418